Green Energy and Technology

For further volumes:
http://www.springer.com/series/8059

Anna Magrini
Editor

Building Refurbishment for Energy Performance

A Global Approach

Springer

Editor
Anna Magrini
Department of Civil and Architecture
Engineering
University of Pavia
Pavia
Italy

ISSN 1865-3529 ISSN 1865-3537 (electronic)
ISBN 978-3-319-37548-9 ISBN 978-3-319-03074-6 (eBook)
DOI 10.1007/978-3-319-03074-6
Springer Cham Heidelberg New York Dordrecht London

Printed on acid-free paper

Springer is part of Springer Science+Business Media (www.springer.com)

Contents

Abstract

The book reports on guidelines dealing with the most suitable actions for energy saving in buildings refurbishment, according to five different themes: opaque and transparent building envelope, heating systems, solar energy use and methods for the energy performance assessment.

Energy saving measures, for a series of European typical wall structures are defined considering an integrated approach that takes into account thermal and hygrometric performances.

The proposal and the evaluation of different strategies also regard energy saving by windows substitution and solar shading device application to existing buildings. The evaluation of the effects on energy consumption, thermal comfort, acoustics and lighting of combined solutions of different windows and screens are proposed in a global perspective evaluation.

A list of suggestions, useful to improve energy efficiency of heating systems, is investigated and supported by calculations of the energy consumptions for the considered solutions. Different geometrical characteristics, climatic conditions and heating systems are taken into account to be useful in a wide field of cases.

The interventions by solar energy for refurbishment applications are described and supported by active systems sizing and by detailed evaluations about the interaction between solar energy and buildings, in order to find the appropriate solutions to use solar radiation.

Finally the importance of accurate energy performance assessment is discussed. The most suitable strategies to refine a building simulation model are described, according to the calibration protocol proposed by ASHRAE Guidelines 14/2002. Moreover, the results of the analyses on real case studies are reported.

The attention is focused on the interaction between building envelope and energy systems, that must be considered carefully, to find the most suitable solutions for reducing energy consumption and improving system efficiency.

Introduction

In Europe the building sector has a strong impact on emissions of greenhouse gases. Existing buildings represent a strategic field for reducing consumptions since they contain the potential energy saving which has to be realized to reach Kyoto Protocol's targets. In fact, they are featured by some critical elements such as lack of insulation, low efficiency hot water production and poor performance of the components of heating systems.

Most of the buildings erected after World War II present poor energy performance. The age of a building often has direct consequences on its conditions, not only because of the lack of services and facilities and the overall obsolescence, but also for the technological choices of the period in which it was constructed.

Therefore, the book deals with efficient methodologies aimed to reduce greenhouse gas emission in the building sector, developed on the basis of the different experiences regarding the building envelope (walls and frames), the heating systems, the use of solar energy and the assessment of the energy sustainability of the proposed solutions.

After a brief introduction to the physical fundamentals involved in the study, results of the investigations are presented in a practical way, to support cost-effective technical strategies by technicians and also by Public Administration, to promote actions for energy saving in the most critical fields and with the most economic advantage. Moreover, the proposed actions could support energy performance assessment, energy labelling and diagnosis of the existing buildings and the application of sustainability protocols.

The presentation of the various aspects regards solutions for building envelope energy saving, related to the following elements:

- Opaque structures: calculation of insulation thicknesses to reduce energy consumption and verification of the risk of condensation in various climatic conditions.
- Transparent structures: efficiency assessment of different windows and solar shading devices for energy saving and for thermal, acoustic and visual comfort improvement.

A series of solutions which allow a considerable reduction of energy consumption is provided, with particular attention to low-cost technologies.

The interventions on the building envelope (walls/windows) must be considered in connection to the heating system improving efficiency, if possible, by means of the use of solar energy.

The theme regarding heating installations starts with the identification of the effective solutions in terms of energy saving. The systematic study will consider adequately different housing types, various climatic conditions and the rapid progress in the heating system technologies field.

The exploitation of solar energy for the reduction of energy requirements of buildings needs greater clarity on the contribution that the technology can offer. The objectives provide assessments on the estimate of energy obtained by active solar systems (thermal collectors and/or photovoltaic systems), and also studies on passive technologies (solar greenhouse, Trombe wall, solar chimney) and on the corresponding increase of environmental sustainability. Moreover, the solar radiation action will be analysed considering its effects on building components (glass and walls) in dynamic conditions.

To complete the analyses, the performance evaluation of buildings and the potential of interventions will be evaluated on the basis of global energy consumption by means of dynamic simulation models.

The chapters contain a collection of technical solutions supported by the research results. They are realised to represent globally a useful basis of knowledge to support the technical choices for the energy performance renovation of residential buildings.

Opaque Building Envelope

A. Magrini, L. Magnani and R. Pernetti

Abstract The refurbishment of building opaque envelope represents an important approach for the reduction in global European energy consumption as prescribed by the Directive 2010/31/EU. Therefore, the knowledge of existing housing stock and their hygrothermal behaviour is necessary to define effective energy conservation measures. In this chapter, starting from the definition of the main thermal and hygrometrical parameters, a systematic analysis of the most common wall typologies in European construction is carried out for the definition of a structure catalogue. Moreover, for a set of envelope insulation interventions, the hygrothermal behaviour is evaluated and some diagrams for the assessment of the optimal intervention are defined. Finally, the incidence in terms of energy performance is tested considering a case study.

Nomenclature

A_f	Internal floor area of the conditioned space (m^2)
EP_{gl}	Global energy performance index ($EP_H + EP_w$) [kWh/(m^2 year)]
$EP_{H,env}$	Energy performance index in the heating season for building envelope [kWh/(m^2 year)]
EP_H	Energy performance index in the heating season [kWh/(m^2 year)]
f_a	Decrement factor (–)
L	Thickness (cm)
M_s	Specific mass (kg/m^2)
$Q_{H,gn}$	Total heat gains for the heating mode (MJ)
$Q_{H,ht}$	Total heat transfer for the heating mode (MJ)
$Q_{H,nd}$	Building energy need for continuous heating (MJ)
Q_{int}	Sum of internal heat gains (MJ)
$Q_{l,e}$	Emission subsystem thermal losses (MJ)
Q_{sol}	Sum of solar heat gains over the given period (MJ)

A. Magrini (✉) · L. Magnani · R. Pernetti
Department of Civil Engineering and Architecture, University of Pavia, Pavia, Italy
e-mail: magrini@unipv.it

A. Magrini (ed.), *Building Refurbishment for Energy Performance*,
Green Energy and Technology, DOI: 10.1007/978-3-319-03074-6_1,

Q_{tr}	Total heat transfer by transmission (MJ)
Q_{ve}	Total heat transfer by ventilation (MJ)
R	Thermal resistance (m^2K/W)
S	Time shift (h)
U	Thermal transmittance [$W/(m^2\ K)$]
Y_{ie}	Dynamic thermal transmittance [$W/(m^2K)$]
ϕ_{int}	Heat gains from internal heat sources (W)
η_e	Emission subsystem efficiency (–)
$\eta_{H,gn}$	Gain utilization factor (–)
$\eta_{y\lambda}$	Heating system global efficiency (–)
ρ	Density (kg/m^3)
c	Heat capacity [J/(kg K)]
λ	Thermal conductivity [W/(m K)]
κ_i	Areal heat capacity [$kJ/(m^2\ K)$]

1 Introduction

The energy-saving target, in the refurbishment of existing buildings, requires considerable efforts on the choice of appropriate solutions that must be energy-efficient and feasible from a technical and economical point of view. Actions on the building envelope are not always possible, and in any case, they must be evaluated with care: higher thermal insulation can be achieved, for example, causing worse hygrometrical behaviour of the walls. Moreover, under certain conditions, it may be convenient to consider jointly the choices for a good thermal insulation and for a good sound insulation.

The improvement of the energy performance of existing buildings is one of the primary goals of the most recent European Directives, starting from 2002/91/EC.

In the case of renovation or maintenance of building walls, the minimum requirements imposed at national level usually indicate precise limits for the values of walls thermal transmittance. In addition, the problem of indoor over-heating in summer is considered in some countries by imposing limits related to the thermal inertia parameters (periodic thermal transmittance, decrement factor, etc.).

Moreover, in addition to the correct evaluation of the thermal and hygrometric parameters of existing structures, the definition of organic strategies to improve the energy performance of the building envelope may be useful.

The need to follow the regulatory minimum requirements must be blended with the technical feasibility and the most appropriate choice of insulating materials.

The classification and analysis of existing walls is a starting point for the study and the choice of appropriate energy-saving actions, taking into account also the issues related to heat storage and hygrometrical aspects.

In the following, a brief overview on typical building walls around Europe is considered. The calculation method of the energy performance of buildings is resumed, and the significant parameters of the opaque envelope are highlighted. Some definitions and fundamental concepts, on which the calculations to be carried out are based, are briefly summarized. The application of the calculation methods is supported by examples. Some considerations on the choice of insulating materials are proposed, considering both the thermal and hygrometrical behaviours of walls.

2 Residential Building Structures of the Twentieth Century: Database of Walls and Buildings

The analysis of the existing building stock represents the first step to define general actions for improving the quality of constructions. In particular, it is useful to catalogue the main building typologies and their structural features according to the construction period and the geographical area, as reported in Table 1 (COST C16, 2007).

As it is shown in Table 1, it is apparent that in the central-northern part of Europe, the concrete and the precast panels are the main technologies, while in the Mediterranean area the envelope is often made of hollow and massive bricks.

Furthermore, in the past recent years, the aim of some research programmes is the definition of a database of existing construction typologies, which represents a strategic operation aimed to begin a structured energy refurbishment programme for low-performance buildings.

Wider information on the existing structures are useful for calculations of standard interventions, which allow to reach established efficiency levels in a systematic way; therefore, their collection may represent a first step for planning.

Some European projects were devoted to a better knowledge of existing building energy performance and refurbishment. An organized analysis of the existing building stocks (INVESTIMMO 2001–2004) provides for useful information related to the most common deterioration problems of the envelope structures and the general state of conservation of European constructions. The analysis was developed on 300 buildings in six Europe countries (Italy, Denmark, Switzerland, Germany, Greece and France) and it highlights that quite half of the buildings considered had façade thermal insulation damaged. If the results can be projected on a larger scale, the energy saving from the façade's restoration could be significant globally.

Moreover, a more recent database of construction typologies, energy-related properties of structure and energy performance assessment (TABULA, 2009–2012) allows defining effective energy conservation measures.

In particular, it aimed to assess how a set of parameters (i.e. construction year, building size) influences the energy performance of existing buildings and the possibilities of energy saving; this acknowledgement allows to define, at national level, general plans for renovation and energy refurbishment.

Table 1 Main features of the European housing stock

Country	Period	Typology	Walls	Outside finishing
Portugal	1950–1960 1960–1980	Multi-family	Single/double-leaf hollow brick walls	Stone, concrete. Plaster, paints, ceramic tiles
Italy	1949–1956 1957–1980	Building block Building block, tower	Single/double-layer hollow brickwork, precast concrete panels	Plaster, paints
Greece	1949–1980	Building block	Massive/perforate brickwork	Plaster, light-colour paints or ceramic stone tiles
Cyprus	1949–1980	Building block	Single-layer brickwork	Plaster, reflective paints
Netherlands	1946–1960 1960–1980	Building block Building block	Concrete block walls, concrete panels	Brick, plaster, concrete panels
Belgium	1949–1980	Building block, row houses	Concrete panels	Concrete, plaster
France	1949–1974 1975–1989	Building block Building block	Concrete walls, concrete panels	Stone, plaster, concrete panels
Sweden	1950–1960 1960–1980	Building block, tower Building block	Brickwork, precast concrete panels	Concrete, plaster, metal sheet
Denmark	1946–1980	Building block	Massive brick masonry, precast concrete panels, lightweight panels	Facing bricks, fibre-reinforced plaster
Germany	1949–1980 1958–1978	Building block Small multi-family, big multi-family	Massive/perforated brick work, concrete panels	Brick, stone metal or wood coverings, concrete, plaster, paints
Poland	1947–1974	Building block	Precast concrete panels	Plaster, paints, concrete
Hungary	1960–1969 1970–1980	Building block Building block	Precast concrete panels	Concrete

The high thermal transmittance of the building envelope evaluated for a 230 multi-storey buildings sample taken all over Europe confirms the problem of the lack of insulation and the need of refurbishment (Fig. 1, no insulation dots). The *U*-values show a global reduction quite only in the last 10 years.

Furthermore in Fig. 1, the possible reduction in wall thermal transmittance is shown applying an insulation layer, respectively, 2 and 4 cm of mineral wall [$\lambda = 0.036$ W/(m K)]. It is apparent that the thermal transmittance decreases according to the insulation thickness with the best results for the worst cases. Therefore, the most effective actions can be addressed to the thermal insulation of buildings with the lowest energy performances: the results may contribute better to optimize the efforts required by the European Directives.

The scenario analysis of the possible energy savings through the improving of building energy performance represents a target of a more general European

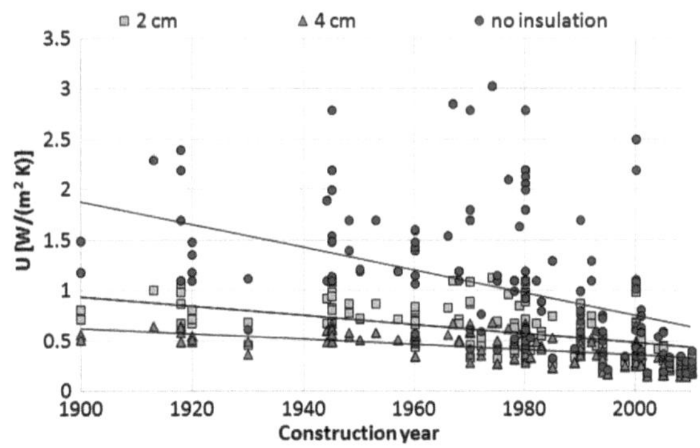

Fig. 1 Thermal transmittance U related to building construction year

refurbishment planning. In particular, the project EPISCOPE (2013–2016), based on the national residential building typologies developed during the TABULA project, is planned to track the energy refurbishment progress of housing stock at different scales. Various refurbishment measures will be determined and compared with those activities needed to attain the relevant climate protection targets.

The preliminary analysis of TABULA proposed two refurbishment levels, to enhance the energy saving of the actual existing building (Level 1):

- Level 2: standard refurbishment measures, commonly used in the countries.
- Level 3: advanced refurbishment measures that use the best available technologies.

In Fig. 2, transmission losses of a series of multi-familiar houses, built in the twentieth century (most of all after the Second World War, until 1995), are shown, referring to the existing conditions (Level 1) and the two levels of interventions (2 and 3).

It is apparent, also from these calculations, that the energy refurbishment of high-loss buildings, both because of the climate conditions and because of the level insulation of the actual envelope, is associated to high energy savings. On the other hand, increasing the energy performance of low-losses buildings is harder, even adopting advanced refurbishment techniques. Anyway, only in a few cases, the annual transmission losses account for less than 50 kWh/(m^2 year) although the thermal transmittance of the envelope is reduced to 0.07–0.3 W/(m^2 K).

3 Building Energy Performance: The Envelope Influence Calculation by Means of a Quasi-Steady-State Approach

The EN ISO 13790 Standard (Energy performance of buildings—Calculation of energy use for spaces—heating and cooling) provides, at European level, some indications on the calculation methods for the design and evaluation of thermal and energy performance of buildings.

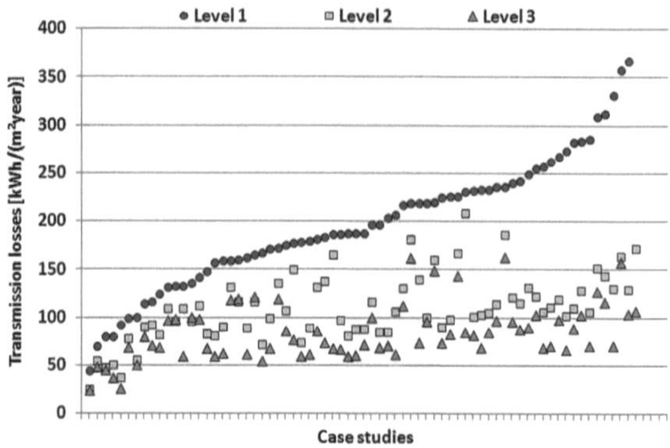

Fig. 2 Annual transmission losses according to the existing situation (*Level 1*) and intervention (*Levels 2* and *3*)

Two approaches are possible:

- quasi-steady-state methods, calculating the heat balance over 1 month or the whole season, taking into account dynamic effects by the simplified determination of a gain utilization factor;
- dynamic methods, calculating the heat balance over 1 h and taking into account the heat stored and released from the mass of the building in a detailed way.

The building energy need for space heating, $Q_{\mathrm{H,nd}}$, in conditions of continuous heating, is calculated by:

$$Q_{\mathrm{H,nd}} = Q_{\mathrm{H,ht}} - \eta_{\mathrm{H,gn}} Q_{\mathrm{H,gn}} \tag{1}$$

where

- $Q_{\mathrm{H,nd}}$ is the building energy need for continuous heating, assumed to be greater than or equal to 0 (MJ);
- $Q_{\mathrm{H,ht}}$ is the total heat transfer for the heating mode (MJ);
- $Q_{\mathrm{H,gn}}$ gives the total heat gains for the heating mode (MJ);
- $\eta_{\mathrm{H,gn}}$ is the dimensionless gain utilization factorand the total heat transfer, $Q_{\mathrm{H,ht}}$, is given by

$$Q_{\mathrm{H,ht}} = Q_{\mathrm{tr}} + Q_{\mathrm{ve}} \tag{2}$$

where Q_{ve} is the total heat transfer by ventilation and Q_{tr} is the total heat transfer by transmission, which depends on temperature difference between indoor and outdoor environment $(\theta_i - \theta_e)$, on the heating or cooling time period *t* and on H_{tr}, the overall heat transfer coefficient by transmission, determined by

$$H_{\mathrm{tr}} = H_{\mathrm{D}} + H_g + H_U + H_A \tag{3}$$

where

- H_{D} direct heat transfer coefficient by transmission to the external environment (W/K);
- H_{g} steady-state heat transfer coefficient by transmission to the ground (W/K);

- H_U transmission heat transfer coefficient by transmission through unconditioned spaces (W/K);
- H_A heat transfer coefficient by transmission to adjacent buildings (W/K).

In general, H_x, representing H_D, H_g, H_U, or H_A, consists of three terms (see EN ISO 13789):

$$H_x = b_{tr,x}\left[\sum_i A_i \cdot U_i + \sum_k l_k \cdot \psi_k + \sum_j \chi_j\right] \tag{4}$$

where

- A_i area of the i-element of the building envelope (m^2);
- U_i thermal transmittance of the i-element of the building envelope [W/(m^2 K)];
- l_k length of the k-linear thermal bridge (m);
- ψ_k linear thermal transmittance of the k-thermal bridge [W/(m K)];
- χ_j point thermal transmittance of the j-point thermal bridge [W/(K)];
- $b_{tr,x}$ adjustment factor for the external temperature.

The adjustment factor $b_{tr,x}$ has to be applied when the envelope element borders on a space which has a different temperature than external environment. As established by the EN ISO 13790, for the monthly and seasonal applications, the total heat transfer by transmission, Q_{tr}, is calculated for each month or season and for each zone and depends on the overall heat transfer coefficient by transmission and on the monthly mean external and internal temperatures.

4 Main Parameters for the Thermal Characterization of Walls

As seen in the previous section, the evaluation of the envelope thermal losses requires the assessment of the heat transfer coefficient H_{tr}, which depends on the thermal transmittance of the building walls and thermal bridges.

4.1 Building Walls: Thermal Transmittance

In order to define the thermal transmittance of walls, different methodologies could be applied as shown in Fig. 3.

The thermal transmittance evaluation for new construction buildings represents an easy operation, since the thermo-physical properties are reported in the technical documents of the project. On the other hand, for existing constructions, the data related to the thermal behaviour of materials are usually not easily available.

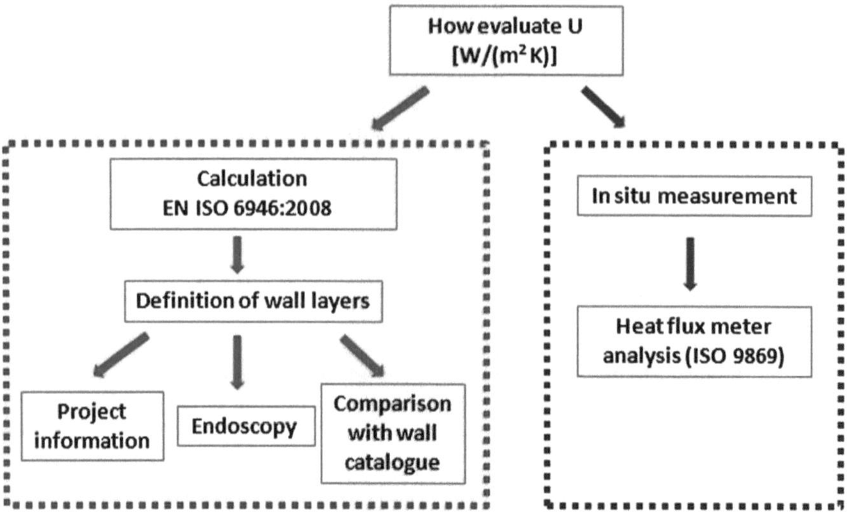

Fig. 3 Wall *U*-value definition

Surface heat transfer coefficients and air layer thermal resistances can be calculated according to the values defined by the EN ISO 6946 Standard, which reports all the indications for the evaluation of thermal transmittance.

4.1.1 Example of Thermal Transmittance Evaluation

Given a wall structure with its layers' characteristics and thermal properties, it is possible to calculate the thermal transmittance. In Table 2, the calculation details are reported, considering that:

- the thermal resistance of homogeneous layers can be calculated as: $R = L/\lambda$
- the thermal transmittance U is obtained as: $U = 1/R_t$.

4.2 Thermal Bridges Evaluation

In the envelope energy performance assessment, thermal bridges play an important role because they influence the heat flow rate and the internal surface temperature, as indicated in the Eq. 4 for the heat transfer coefficient.

Thermal bridge is defined as follows: *part of the building envelope, where the otherwise uniform thermal resistance is significantly changed by full or partial penetration of the building envelope by materials with a different thermal conductivity, and/or a change in thickness of the fabric, and/or a difference between*

Table 2 Example of thermal transmittance evaluation

Layer	L (cm)	ρ (kg/m^3)	λ [W/(m K)]	R (m^{2}K/W)
Internal surface resistance				0.13
1 Internal plaster	2	1,400	0.700	0.029
2 Hollow bricks[a]	8	800	–	0.200
3 Air layer[b]	2.5/30	1.2	–	0.180
4 Concrete blocks	20	1,400	0.500	0.400
External surface resistance				0.04
Total resistance R_t (m^2 K/W)				0.979
Thermal transmittance $U = R_t^{-1}$ [W/(m^2 K)]				1.021

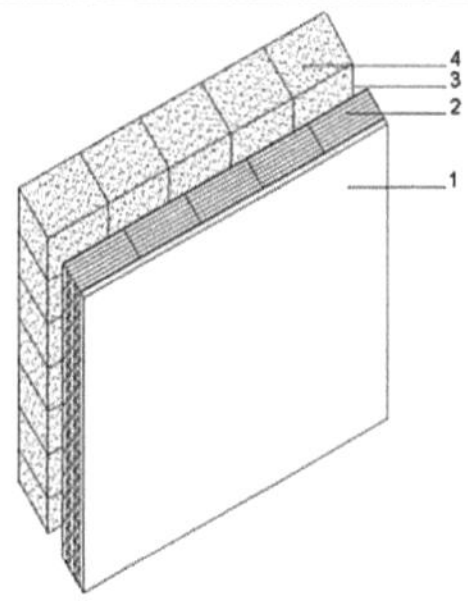

[a] Equivalent thermal resistance for hollow bricks
[b] Values for unventilated air layers from the EN ISO 6946 Standard

internal and external areas, such as occur at wall/floor/ceiling junctions (EN ISO 10211 Standard).

In general, they are located in any connection between building components or where the building structure changes composition, and in these elements, the hypothesis of mono-dimensional flow is not correct, since they cause two-dimensional or three-dimensional heat flows.

In order to take into account the incidence of a thermal bridge on the whole heat flow, the linear thermal transmittance ψ [W/(m K)] is used; it represents the heat flow rate in steady-state conditions, divided by the length of the junction and by the temperature difference between internal and external surfaces.

In particular, reference values could be used for some standard structures referring to a catalogue of thermal bridges and values of ψ (EN ISO 14683) in relation to some different geometrical dimension, as follows:

- ψ_e: linear thermal transmittance determined according to the external dimensions and measured between the finished external faces of the building external elements;
- ψ_{oi} linear thermal transmittance determined according to the overall internal dimensions and measured between the finished internal faces of the building external elements including the thickness of internal partitions;
- ψ_i: linear thermal transmittance determined according to the internal dimensions and measured between the finished internal faces of each room in a building excluding the thickness of internal partitions.

Furthermore, if the analysed thermal bridge is not included in the standard catalogue, it is possible to determine its linear thermal transmittance through numerical methods. The EN ISO 10211 Standard provides for the definition of a geometrical model of a thermal bridge for the numerical calculation of heat flows and surface temperatures considering the following assumptions:

- all physical properties are independent from temperature;
- there are no heat sources within the building element.

Moreover, a thermal coupling coefficient is defined depending both on the heat flows and on the linear thermal transmittance.

In particular, given the geometrical features and the thermal properties of the thermal bridge, it is possible to use a numerical model, whose output is the global heat flows through the element. The coupling coefficient can be determined by the following equation:

$$\varphi_1 = L_{2D} \cdot (\theta_i - \theta_e) \tag{5}$$

where

- θ_i and θ_e represent the internal and the external air temperature;
- φ_1 is the linear thermal flow of the thermal bridge;
- L_{2D} is the coupling thermal coefficient of the thermal bridge, which has to be evaluated

Then, the linear thermal transmittance can be assessed through:

$$\psi = L_{2D} - \sum_{j=1}^{Nj} U_i l_j \tag{6}$$

where

- U_i is the thermal transmittance of one-dimensional component separating the two environments;
- l_j is the length within the 2D geometrical model over which the value U_j is applied;
- N_j is the number of one-dimensional components.

4.3 Dynamic Thermal Characteristics of the Envelope

The evaluation of the dynamic thermal behaviour of the envelope is based on the definition of transient building simulation models that consider the heat storage in the structure and the time dependence of boundary conditions.

These models should be adopted for the analysis of summer behaviour; in fact, for the heating season, the evaluation by quasi-steady-state methods allows to obtain reliable results. Nevertheless, the quasi-steady-state methods do not represent accurately the thermal behaviour of buildings during the cooling season; therefore, a transient model has to be applied to have accurate results.

Moreover, the European Directive 2020/31/EU highlights that the cooling energy needs have been increased during the last years, and it suggests to focus on strategies which enhance the thermal performances also during summer period, and on the thermal properties that influence the internal overheating, such as heat thermal capacity, specific mass, periodic thermal transmittance, time shift and decrement factor. In particular, the Directive establishes that minimum requirements according to the climatic conditions have to be fixed at national level; therefore, for the countries of the Mediterranean area, dynamic envelope features that reduce the cooling energy needs have to be evaluated.

The EN ISO 13786 Standard defines the dynamic thermal characteristics of the envelope.

One of the significant parameters in this field is the periodic thermal transmittance Y_{ie} [W/(m^2 K)], which is defined as the amplitude of the heat flow rate through the component surface adjacent to a zone which is kept at constant temperature, divided by the amplitude of the temperature in the adjacent zone (which also could be the external environment).

Another parameter, which represents the effect of the envelope on the inward heat flow rate, is the decrement factor, which is expressed by the ratio between the periodic thermal transmittance Y_{ie} [W/(m^2 K)] and the quasi-steady-state thermal transmittance U [W/(m^2 K)]:

$$f = \frac{|Y_{ie}|}{U} \tag{7}$$

The envelope influence on the thermal behaviour is also represented by the time shift Δt, which is defined as the period of time between the maximum amplitude of a cause and the maximum amplitude of its effect.

5 Vapour Transmission Problems

The knowledge of moisture transfer mechanisms, material properties, initial and boundary conditions is often limited.

The building structures are exposed to variable internal and external climatic conditions, as they are usually influenced by outdoor air temperature, relative humidity, solar radiation, rain or are in contact with humid soil. In addition, some building components (concrete, etc.) in the construction process are characterized by high moisture contents that should be dried in short times to avoid the risk of damage to the other components.

The moisture problems that may affect building structures are various: capillary rise of water in the walls, condensation inside building components due to infiltration of indoor air (hot and humid), problems with tightness to rainwater, salts migration inside materials, etc. and, not less important phenomena, hygrometric surface problems (growth of mould and moisture condensation) and water vapour condensation inside structures.

Therefore, the hygrometric assessment of existing building components presents considerable practical interest and it can be used, for example, to determine whether the operating conditions may lead to a progressive deterioration of the structures.

The constructions can indeed present different degradation levels depending on the types of aggression to which they are subjected: one of the most significant causes is represented by the presence of moisture or liquid water. Some materials, in fact, absorbing water vapour, increase their volume and can generate movements and internal tensions to the structures causing leakages and deformations.

Moreover, the presence of liquid water can cause an increase of thermal conductivity, thus degrading the heat transmission properties. The elimination of the serious damage resulting from the presence of water in the structures may represent a significant item of the costs for the building refurbishment.

The moisture transport and condensation in porous media is very complex, so even today, the degree of knowledge of the interaction between the various transport mechanisms and properties of building materials is unsatisfactory, incomplete and under development.

The International Standard, that gives calculation guidelines to assess of surface mould growth and the interstitial condensation for the prevention of these phenomena, adopts therefore simplified methods. The results of the calculations should be considered an approximation of more complex physical phenomena and therefore characterized by uncertainty. The standardization of simplified methods obviously does not preclude the use of more accurate calculation methods.

The following considered aspects are related only to a part of the hygrometric problems that afflict building structures: in fact, the surface mould growth and the interstitial condensation are considered essentially due to the vapour transport through the wall caused by a difference in partial vapour pressure of the internal/external environment.

The calculations follow the indications provided by the most recent release of the EN ISO 13788 Standard. Therefore, the moisture transfer through a wall is considered purely a function of inside and outside temperature and humidity conditions, and of dimensional and thermo-physical characteristics of its layers.

The diffusion of water vapour through a structure by itself may not cause problems, if it does not cross areas with too low temperature to cause condensation. In particular, two different phenomena can be considered:

- at the wall surface: high values of relative humidity or even, vapour condensation (surface phenomena);
- inside the material layers: vapour condensation as the vapour pressure reaches the saturation pressure (interstitial phenomena).

These phenomena can occur at different times and the distinction between them may not always be clearly defined: the effects of interstitial condensation may arise on the surface and this can happen even after a long time and in zones far from the ones where it took place.

Side effects resulting from these phenomena can produce degradation of buildings and unhealthy environments in the following forms:

- growth of fungal colonies on the inner surface of the building envelope;
- the presence of condensed water on the surface and inside of the walls;
- decay of wooden structures;
- degradation of plaster;
- reduction in the thermal insulation;
- dimensional changes and damage of artefacts;
- migration of salts, efflorescence.

The occurrence of these phenomena, as well as depending on the characteristics of the building and the external climatic conditions, increases with the indoor vapour production and with the reduced air renewal.

5.1 Vapour Diffusion: Basic Parameters

To address the problem of the vapour transfer in walls it is necessary to consider the mechanisms that produce its movement.

Mainly, the vapour flow within a material is determined by a difference between the partial vapour pressures or the vapour concentrations in the air in contact with the two faces of the wall. This phenomenon can be observed commonly in building structures.

Human activities modify the vapour concentration in the internal environment that, therefore, especially in winter, is higher than the one in the external environment. It can be assumed that, at least in the winter season, the moisture transfer is from inside to outside the building.

The thermo-physical parameters of the materials involved in a simplified approach to the problem, in steady-state conditions, are mainly the thermal conductivity λ [W/(m K)], which affects the temperature profile within the layers, and the water vapour permeability δ [kg/(m s Pa)], which determines the greater or lesser capacity of a material to be crossed by the water vapour. For some materials, it is indicated the hygroscopic resistance factor $\mu = \delta_o/\delta$ (–), where δ_o is the water vapour permeability of the air, which assumes, as the reference value, 200×10^{-12} kg/(m s Pa).

To describe in a quantitative way the mass transport by diffusion, the Fick's law can be applied: it relates the vapour flow rate per unit area g'_v through a material (thickness L and vapour permeability δ), with the difference ΔP_v between the vapour pressures of the air on its surfaces. Assuming steady-state conditions, g'_v [kg/(m^2 s)] can be expressed as

$$g'_v = \delta \Delta P_v / L \tag{8}$$

It remains constant if there are not condensation phenomena in the internal layers. They may occur if the internal temperatures are low enough to cause low values of saturation vapour pressure: local vapour pressure can reach easily the maximum value of saturation.

The expression often refers to the equivalent thickness s_d [m]:

$$s_d = L\delta_o/\delta = L\mu \tag{9}$$

where

$\mu = \delta_o/\delta$ Hygroscopic resistance factor

δ_o Vapour permeability of air = 193×10^{-12} kg/(m s Pa), approximated in the calculations to 200×10^{-12} kg/(m s Pa)

Therefore, the vapour flow rate g'_v [kg/(m^2 s)] can be written as

$$g'_v = \delta \Delta P_v / L = \delta_o \Delta P_v / s_d \tag{10}$$

5.2 *The Assessment of the Risk of Interstitial Condensation Due to Water Vapour Diffusion*

The Glaser method allows to verify the risk of interstitial condensation in a building wall through a graphical representation.

It compares the vapour pressure values with the saturation pressure ones, calculated as a function of the temperature distribution within the different layers of the wall. From the graphical comparison of the two trends, it is possible to put in evidence the condition in which there is condensation of the vapour, which occurs if the vapour pressure reaches the value of the saturation.

The wall passes the test if two criteria are met:

- the calculation (usually on a monthly basis) demonstrates that any condensation can be completely dried throughout the year.
- the condensation in a layer does not exceed the limit values of the materials involved.

Although the Glaser method is the reference for the evaluation of the vapour condensation risk in building structures, the method has limitations that can lead to mistakes.

Indeed, the thermo-physical characteristics of the materials, such as thermal conductivity, depend on the moisture content of the materials, which is influenced by the capillary absorption, by the liquid water transfer, or even by the movements of air through cavities or by the superficial absorption of rain.

The method considers the built-in water dried out and it neglects some effects:

- the variation of material properties with moisture content;
- the capillary suction and the liquid water transport within the material;
- the air movement through cavities or air spaces;
- the hygroscopic moisture capacity of materials.

The evaporation and condensation processes that may occur in the materials also involve thermal energy exchanges that modify the temperatures distribution and the corresponding saturation pressures. Moreover, the effects of the solar radiation can contribute to change the distribution of the temperature in the outer layers.

In addition, neglecting the effects listed above, the analysis based on the Glaser method is carried out considering the mono-dimensional moisture transport and assuming constant boundary conditions.

The interstitial condensation assessment is developed referring to the monthly mean values of the climatic data. The verification process can be summarized in the following steps.

Step 1 Boundary conditions
(a) Material properties and geometrical dimensions (thickness).

Thermal conductivity or thermal resistance or conductance and vapour permeability of the layers are the main information needed.

For new buildings, the necessary data can be found in the technical specifications, while for existing buildings, some indications provided by international/national standards can be considered as reference values. For example, EN ISO 10456:2008 "Materials and products for buildings—Hygrothermal properties, tabulated design values and procedures for determining declared and design thermal values" can be a reference for some materials.

Some indications (by EN ISO 13788) should be taken into account:

- elements with a thermal resistance greater than 0.25 m^2 K/W should be subdivided into a number of notional layers, each one with thermal resistance $\leq$0.25 m^2 K/W; these subdivisions are treated as separate material layers with interfaces between them in all calculations;
- if the wall presents a ventilated cavity, the presence of materials between this and the outside is ignored;
- some materials, such as aluminium foil, effectively prevent vapour transmission: their water vapour resistance factor is practically infinite. However, since for the calculation procedure is required a finite value, for these materials it is assumed $\mu = 100{,}000$. This can lead to predict small quantities of condensed water that should be neglected.

Table 3 Surface thermal resistance for the interstitial condensation calculations

Heat flow direction	Internal thermal resistance R_i (m^2 K/W)
Upwards	0.10
Horizontal	0.13
Downwards	0.17
All	External thermal resistance R_e (m^2 K/W) 0.04

(b) Surface thermal resistance. To calculate the surface thermal resistance, the reference standard is EN ISO 6946 (Table 3), as well as for the thermal resistance of cavities. For air layers, the equivalent thickness s_d is considered to be equal to 0.01 m, regardless of their real thickness and position (vertical/horizontal).

(c) External boundary conditions: monthly mean temperature and vapour pressure values are considered.

(d) Internal boundary conditions: air temperature. The indoor air temperature is assumed according to the expected use of the building. For residential buildings, in the absence of specific information, the following values may be considered (ref. Italian regulations):

- $t_i = 20$ °C in the heating period (heating system is on);
- $t_i = 18$ °C out of the heating period, when the monthly average outdoor temperature is $t_e < 18$ °C;
- $t_i = t_e$ when the monthly average outdoor temperature is ≥ 18 °C.

(e) Internal boundary conditions: vapour pressure. If there is a HVAC system, relative humidity and internal temperature are imposed and the corresponding vapour pressure can be calculated. In the absence of a relative humidity control system, the data should be obtained by a hygrometrical balance of the environment.

Alternatively, for maritime-mild climates, in the absence of detailed information on the building use, and of a RH control system, the information provided by the graph in Fig. 4 may be useful.

The horizontal axis represents the external monthly mean temperature. Month-to-month, the difference between internal and external vapour pressure ΔP_v can be determined, rising up from the temperature to meet the upper line of the chosen class. The upper lines are the reference for each class. In principle, residential buildings refer to class 3.

Step 2 Starting month

The climatic conditions of the starting month correspond to the first month in which condensation occurs (if there is). In principle, as a first attempt, the climatic conditions of a month at the end of the autumn can be considered. Once the starting month is individuated, the monthly mean external climatic conditions (temperature and relative humidity or vapour pressure) are used to calculate the amount of condensed/evaporated water (if any) month by month to complete the annual cycle.

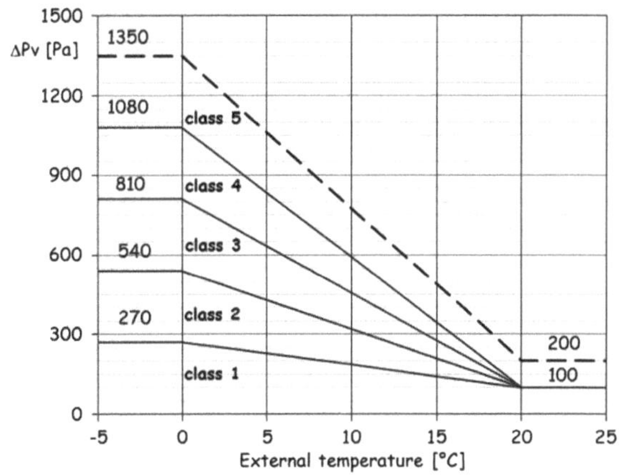

Fig. 4 $P_{vi} - P_{ve}$ as function of humidity classes. Legend: *Class 1* unoccupied buildings, storage of dry goods. *Class 2* offices, dwellings with normal occupancy and ventilation. *Class 3* buildings with unknown occupancy. *Class 4* Sports halls, kitchens, canteens. *Class 5* Special buildings, e.g. laundry, brewery, swimming pool

Step 3 Calculation of the temperature and saturation pressure distribution inside the wall.

All data are referred to interfaces between layers. The temperatures at each interface between layers are calculated, as function of the thermal resistances and the temperatures of the indoor–outdoor environments.

The saturation pressure P_s (Pa) is calculated as a function of temperature. The following expressions can be used:

$$t \geq 0\,°\mathrm{C} \qquad P_s = 610.5\, e^{\frac{17.269 \cdot t}{237.3+t}} \tag{11}$$

$$t < 0\,°\mathrm{C} \qquad P_s = 610.5\, e^{\frac{21.875 \cdot t}{265.5+t}} \tag{12}$$

Step 4 Vapour pressure distribution. Graphical comparison between vapour pressure and saturation pressure values, as a function of the equivalent thickness s_d, allows to verify whether there is condensation. It occurs if the vapour pressure reaches the saturation pressure.

Step 5 In case of the presence of condensation in the trial month conditions: the calculation must be performed for the external climatic conditions of the previous month. The procedure is repeated backwards until it is individuated the first month in which no condensation occurs. Starting from this one, the amount of condensation the next months is calculated.

Step 6 In case of no condensation in the wall: the wall is checked with the climatic conditions of the following month. If there is no condensation in all the 12 months of the year, the structure is free of interstitial condensation. If condensation occurs in a following month, this is considered the starting month for the calculation of the amount of condensation.

Step 7 The presence of condensation during some months. The condensation can evaporate in some other months of the year. In the first month in which the calculation does not show condensation, it is assumed $P_v = P_s$ in the wet layer of the previous month, to calculate the amount of evaporated water. In the following months, the same principle is applied, until all the condensation is progressively dried.

Step 8 Add up all positive (condensation) and negative (evaporation) quantities and verify that the sum of the positive contributions can be balanced by the negative ones over the year.

Step 9 The total amount of accumulated condensate is compared with the maximum values allowed for the different types of materials.

In the EN ISO 13788 standard, the risk of run-off from non-absorbent materials is considered very high over 200 g/m^2 of condensate. In other cases (i.e. Italian regulations), for example, the maximum amount of condensate accumulated in a layer is $M_{c,amm} = 500\ \text{g/m}^2$. For some materials, lower values are taken into account, in function of thickness L (m), density ρ (kg/m^3) and conductivity λ [W/(m K)]:

- Wood $(\rho = 500 - 800\ \text{kg/m}^3) : M_{c,amm} \leq 30\,\rho\,\text{L}$
- Plaster and mortar $(\rho = 600 - 2{,}000\ \text{kg/m}^3) : M_{c,amm} \leq 30\ \rho\ \text{L}$
- Mineral wool $(\rho = 10 - 150\ \text{kg/m}^3) : M_{c,amm} \leq 5{,}000\ \rho\ \text{L}\ [\lambda/(1 - 1{,}7\,\lambda)]$
- Cellular plastic materials $(\rho = 10 - 80\,\text{kg/m}^3) : M_{c,amm} \leq 5{,}000\ \rho\,L\,[\lambda/(1\ -\ 1{,}7\,\lambda)]$

Step 10 The assessment is positive if both conditions referred to in Steps 8 and 9 are satisfied.

5.2.1 Example: Interstitial Condensation Calculation

To illustrate the calculation procedure for the verification of interstitial condensation problems, a wall with the geometrical and thermal characteristics described in the following has been considered.

Step 1 (1a, 1b)—the geometrical structure and thermal parameters are summarized in Table 4. The climatic conditions (1c, 1d) are reported in Table 5. The indoor vapour production (1e) is assumed corresponding to class 3 (Fig. 4); therefore, the internal vapour pressure can be calculated by means of the slope of the corresponding lane, analytically expressed as:

$$P_{vi} = P_{ve} + 100 + [(810 - 100)/20]\,(20\ -\ t_e) \tag{13}$$

The P_{vi} values are resumed in Table 5, as well as the internal temperature of the environment, based on the criteria reported in Step 1 of the procedure. For each layer is calculated also the equivalent thickness for vapour diffusion $s_d = L\delta_o/\delta(\text{m})$ (Table 6).

Step 2 The calculations start with the conditions of the trial month: it is established (arbitrarily) to be October. The corresponding monthly mean values of temperature and vapour pressure are considered. Depending on the

Table 4 Thermal properties of the wall

Concrete insulated precast wall

Layer	L (cm)	ρ (kg/m³)	c [J/(kg K)]	λ [W/(m K)]	R (m² K/W)
1 Internal plaster	1	1,400	1,000	0.700	0.014
2 Concrete panel	1	1,400	1,000	0.580	0.017
3 Mineral wall panel	3	30	670	0.040	0.750
4 Concrete panel	10	1,400	1,000	0.580	0.172
5 External plaster	2	1,800	1,000	0.900	0.022
U [W/(m² K)]	0.873			R_i (m²K/W)	0.13
M_s (kg/m²)	205			R_e (m²K/W)	0.04
κ_i [kJ/(m² K)]	33.1			R_{tot} (m²K/W)	1.146
Y_{ie} [W/(m²K)]	0.555				
S (h)	5.35				
f_a	0.635				

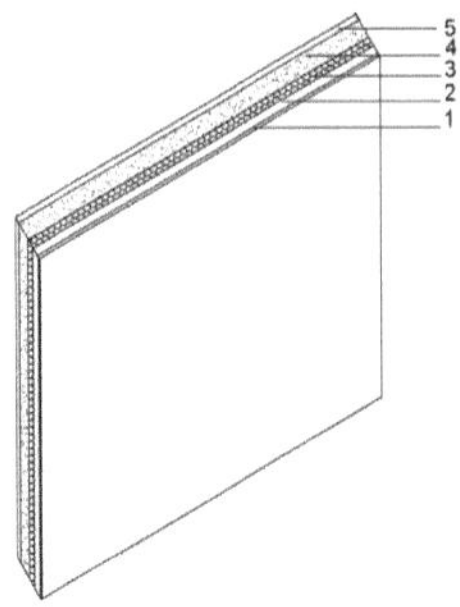

Table 5 Internal and external climatic conditions

	Jan	Feb	Mar	Apr	May	Jun	Jul	Aug	Sep	Oct	Nov	Dec
t_e (°C)	1.5	4.2	9.3	13.5	17.7	22.0	24.4	23.7	19.9	14.0	7.8	3.5
P_{ve} (Pa)	661	685	783	1,046	1467	1,625	1,910	2,036	1622	1,246	896	582
P_{vi} (Pa)	1,418	1,346	1,263	1,377	1,649	1,654	1,854	2,005	1,726	1,559	1,429	1,268
t_i (°C)	20	20	20	20	18	22	24.4	24	19.9	20	20	20

climatic conditions, it is possible that condensation will not occur in October and it will not be necessary to proceed backwards with the search of the first month of the condensation.

Step 3 The temperatures at different interfaces are calculated, based on the outside temperature and on the relationship:

$$t_x = t_{x-1} - (t_i - t_e)R_x/R_T \tag{14}$$

and they are applied for each x-interface:

Table 6 Calculation details for the wall

Layer	L	λ	R	δ	s_d	ΣR_i	Σs_d	Interf
Internal thermal resistance R_i			0.13		0.00	0.130	0	0–1
1 Internal plaster	1	0.700	0.014	18	0.11	0.144	0.11	1–2
2 Concrete panel	1	0.580	0.017	4	0.50	0.161	0.61	2–3a
3a Mineral wall panel	1	0.040	0.250	150	0.013	0.411	0.62	3a–3b
3b Mineral wall panel	1	0.040	0.250	150	0.013	0.661	0.64	3b–3c
3c Mineral wall panel	1	0.040	0.250	150	0.013	0.911	0.65	3c–4
4 Concrete panel	10	0.580	0.172	4	5.00	1.084	5.65	4–5
5 External plaster	2	0.900	0.022	18	0.22	1.106	5.87	5–0
External thermal resistance R_e			0.04			1.146		

Note The thermal resistance of the glass fibre layer is higher than the limit value of 0.25 m^2 K/W, and therefore, it is divided into three sublayers, each characterized by a thermal resistance of 0.25 m^2 K/W

$$t_1 = t_0 - (t_i - t_e) R_1 / R_T = 20 - 0.13(20 - 14)/1.146 = 19.3\,°\mathrm{C}$$
$$t_2 = t_1 - (t_i - t_e) R_2 / R_T = 19.3 - 0.014(20 - 14)/1.146 = 19.2\,°\mathrm{C}$$

and so on for all the layers. All the values are reported in the third column of Table 7. From them, it is possible to calculate the corresponding values of the saturation pressure.

Step 4 To calculate the vapour pressure in each interface, the vapour resistance surface coefficients are neglected, and therefore, the vapour pressures on the two wall surfaces coincide with the corresponding environmental ones:

$$P_{\mathrm{vi}} = P_{\mathrm{v1}} \text{ and } P_{\mathrm{ve}} = P_{\mathrm{v5}} \tag{15}$$

Starting from the assumption that there is not condensation and therefore the vapour flow rate through the wall is constant, the two surface values can be connected by a straight-line segment in a plot P_v versus s_d. Analytically, the vapour pressure of each interface is calculated according to Fick's law:

$$g_{\mathrm{v}}' = \delta_o(P_{\mathrm{vi}} - P_{\mathrm{ve}})/s_{\mathrm{dtot}} = \delta_o(P_{v1} - P_{v2})/s_{\mathrm{d1-2}} \tag{16}$$

and therefore

$$P_{v2} = P_{v1} - s_{\mathrm{d1}}(P_{\mathrm{vi}} - P_{\mathrm{ve}})/s_{\mathrm{dtot}} = 1{,}459 - 0.11(1{,}459 - 1{,}246)/5.87$$
$$= 1{,}455\ \mathrm{Pa}$$

and similarly, applying the same relation to all the following layers:

$$P_{\mathrm{vx}} = P_{\mathrm{v.x-1}} - s_{\mathrm{d.x}}(P_{\mathrm{vi}} - P_{\mathrm{ve}})/s_{\mathrm{dtot}} \tag{17}$$

from which

$$P_{v3} = P_{v2} - s_{\mathrm{d2}}(P_{\mathrm{vi}} - P_{\mathrm{ve}})/s_{\mathrm{dtot}} = 1{,}455 - 0.5(1{,}459 - 1{,}246)/5.87$$
$$= 1{,}437\ \mathrm{Pa}$$

Table 7 a. Temperature, saturation and vapour pressure in the wall (October–January); **b** Temperature, saturation and vapour pressure in the wall (February–April)

Interf	R	s_d	October			November			December			January		
			t	P_s	P_v	T	P_s	P_v	t	P_s	P_v	t	P_s	P_v
(a)														
			20	2,337	1,559	20	2,337	1,429	20	2,337	1,268	20	2,337	1,418
0–1	0.13	0	19.3	2,240	1,559	18.6	2,144	1,429	18.1	2,080	1,268	17.9	2,050	1,418
1–2	0.014	0.11	19.2	2230	1,553	18.5	2124	1,419	17.9	2,053	1,255	17.7	2,021	1,403
2–3a	0.017	0.5	19.2	2217	1,526	18.3	2100	1,374	17.7	2,021	1,196	17.4	1,986	1,339
3a–3b	0.250	0.013	17.8	2043	1,526	15.6	1774	1,372	14.1	1,606	1,195	13.4	1,532	1,337
3b–3c	0.250	0.013	16.5	1881	1,525	13.0	1493	1,371	10.5	1,267	1,193	9.3	1,173	1,336
3c–4	0.250	0.013	15.2	1730	1,524	10.3	1252	1,370	6.9	993	1,192	5.3	890	1,334
4–5	0.172	5.0	14.3	1632	1,258	8.5	1106	916	4.4	836	608	2.5	731	690
5–0	0.022	0.22	14.2	1620	1,246	8.2	1089	896	4.1	817	582	2.2	713	661
	0.04		14.0	1598	1,246	7.8	1058	896	3.5	785	582	1.5	680	661

Interf	R	s_d	February			March			April		
			t	P_s	P_v	t	P_s	P_v	t	P_s	P_v
(b)											
			20	2,337	1,346	20	2,337	1263	20	2,337	1,377
0–1	0.13	0	18.2	2,090	1,346	18.8	2,167	1263	19.3	2,232	1,377
1–2	0.014	0.11	18.0	2,064	1,333	18.7	2,149	1254	19.2	2,221	1,370
2–3a	0.017	0.5	17.8	2,034	1,277	18.5	2,128	1213	19.1	2,208	1,342
3a–3b	0.250	0.013	14.3	1,632	1,276	16.2	1,836	1212	17.7	2,020	1,342
3b–3c	0.250	0.013	10.9	1,302	1,274	13.8	1,580	1211	16.2	1,846	1,341
3c–4	0.250	0.013	7.4	1,032	1,273	11.5	1,355	1210	14.8	1,686	1,340
4–5	0.172	5.0	5.1	875	710	9.9	1,218	801	13.9	1,583	1,059
5–0	0.022	0.22	4.8	857	685	9.7	1,201	783	13.7	1,570	1,046
	0.04		4.2	824	685	9.3	1,171	783	13.5	1,547	1,046

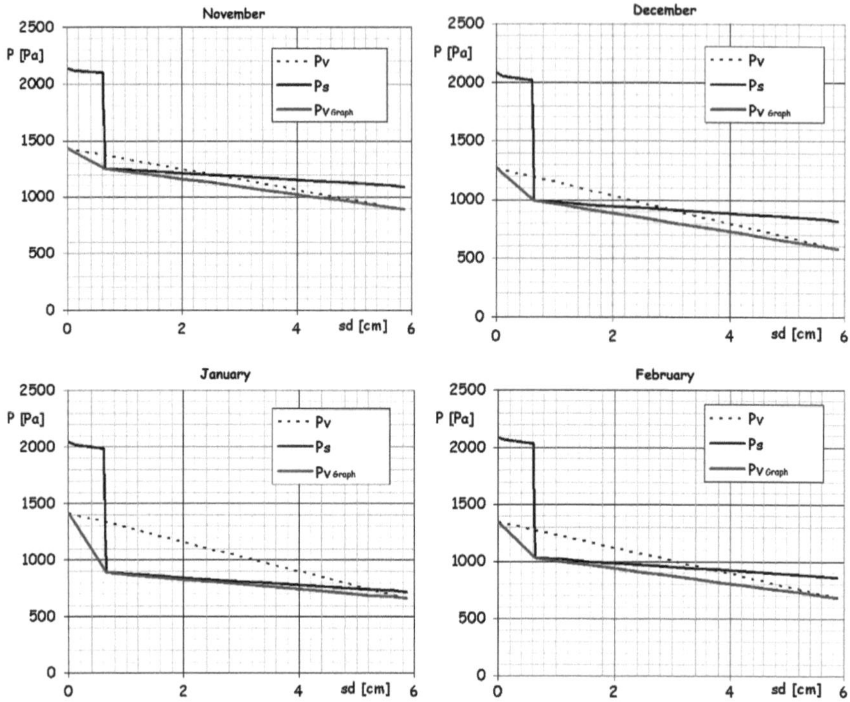

Fig. 5 Glaser diagrams (condensation) for **a** November, **b** December, **c** January, **d** February

and so on for the other layers.

The vapour pressure values in the different interfaces are reported in Table 7.

The calculations for October show that all values of the saturation pressure are higher than the vapour pressure in each interface and therefore no condensation occurs for the conditions of this month. Step 5 can be skipped.

Step 6 The calculations are developed for the 12 months (Table 7) with the input data in Tables 4 and 5. The difference between saturation and vapour pressure in November can be put in evidence: in the interface 3c-4, it can be observed that $P_v > P_s$. The diagrams in Fig. 5 highlight the occurrence.

Then, the month of November is the starting month for the calculation of the amount of condensation that forms at the interface between the two layers.

P_v is plotted as a straight line between the surface values and the P_s lower value (in this case at the interface 3c-4). The difference between the two slopes represents the rate of condensed water.

Step 7 The calculations proceed with the climatic conditions of the subsequent months, to determine the amount of condensate accumulated in the interface for each month. The condition $P_v > P_s$ is observed until February and the same procedure of November is applied. Note that for January, the

Table 8 Amounts of condensation/evaporation quantities

Month	M_c (g/m^2)	Sum (g/m^2)
October	0.0	0.0
November	105.6	105.6
December	183.9	289.5
January	411.1	700.7
February	201.4	902.1
March	−134.9	767.2
April	−309.7	457.5
May	−373.4	84.1
June	−887.9	0.0
July	0.0	0.0

comparison between P_v and P_s shows that $P_v > P_s$, also the interface 3b-3c. From the graph (Fig. 5c), however, it is observed that the minimum P_s corresponds to the interface 3c-4.

For example, the condensate for November is detailed (g_{H_2O} in [kg/(m^2 s)]):

$$g_{H_2O} = g_v{}'_{in} - g_v{}'_{out} = \delta_o(P_{vi} - P_{s3c-4})/(s_{d3c-4}) - \delta_o(P_{s3c-4} - P_{ve})/(s_{dtot} - s_{d3c-4}) \quad (18)$$

$$g_{H_2O} = 200 \times 10^{-12}[(1{,}429 - 1{,}252)/(0.11 + 0.5 + 0.013 + 0.013 + 0.013) + - (1{,}252 - 896)/(5.0 + 0.22)]$$

$$g_{H_2O} = 200 \times 10^{-12}[272 - 68.2] = 4.09 \times 10^{-8}\,\text{kg}/(\text{m}^2\,\text{s}) = 3.53\ \text{g}/(\text{m}^2\,\text{day})$$

The water condensate in the whole month will be

$$M_c = 3.53\ \text{g}/(\text{m}^2\text{day})30\ \text{days} = 106\ \text{g/m}^2$$

The calculation of the amount of condensed water is performed for all months until February: the quantities are indicated in Table 8.

In March, all interfaces present $P_v < P_s$; therefore, the amount of evaporated water is calculated assuming that $P_v = P_s$ in the 3c-4 interface, where the condensate remains concentrated until complete evaporation.

On a graph (Fig. 6), the trends of the saturation and vapour pressure are shown; the mass of evaporated water in the period can be calculated as

$$g_v{}'_{in} = \delta_o(P_{v1} - P_{s3c-4})/(\Sigma s_d)_{1-3c}$$
$$g_v{}'_{in} = 200 \times 10^{-12}(1{,}263 - 1{,}355)/0.65 = -2.83 \times 10^{-8}\,\text{kg}/(\text{m}^2\,\text{s}) \quad (19a)$$

Due to the positive direction of the *x*-axis from inside to outside, the negative sign indicates that the vapour flow is directed from the interface to the indoor

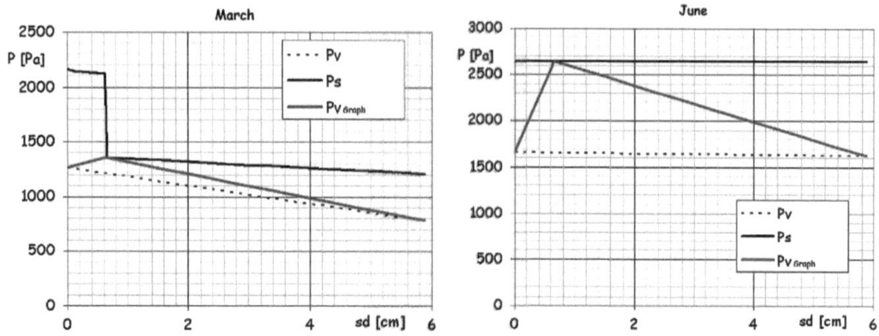

Fig. 6 Example of Glaser diagrams for evaporation: **a** March, **b** June

environment. The vapour flow from the interface 3c -4 to outside can be expressed as

$$\begin{aligned} g_{\mathrm{v}}{}'_{\mathrm{out}} &= \delta_{\mathrm{o}}(P_{\mathrm{s3c-4}} - P_{\mathrm{v4}})/[s_{\mathrm{dtot}} - (\Sigma s_{\mathrm{d}})_{1-3\mathrm{c}}] \\ g_{\mathrm{v}}{}'_{\mathrm{out}} &= 200 \times 10^{-12}(1{,}355 - 783)/(5.87 - 0.65) = 2.19 \times 10^{-8} \end{aligned} \quad (19b)$$

In this case, the positive sign indicates that the vapour flow is directed to outside. The two calculated flow rates show that the vapour transfer occurs from the interface where it condensed in the previous months both to the internal and external environments.

The total amount of evaporated water in March will be

$$\begin{aligned} g'_{\mathrm{H_2O}} &= g_{\mathrm{v}}{}'_{\mathrm{in}} - g_{\mathrm{v}}{}'_{\mathrm{out}} = -2.83 \times 10^{-8} - 2.19 \times 10^{-8} = -5.02 \times 10^{-8}\mathrm{kg/(m^2\,s)} \\ g'_{\mathrm{H_2O}} &= -5.02 \times 10^{-8}\,\mathrm{kg/(m^2\,s)} = -5.02 \times 10^{-8}\,86{,}400 \times 10^{3}\mathrm{g/(m^2\,day)} = -4.34\ \mathrm{g/(m^2\,day)} \\ M_{\mathrm{c}} &= g'_{\mathrm{H_2O}}\,\Delta\tau = [-4.34.0\ \mathrm{g/(m^2\,day)}]\ 31\ \mathrm{days} = -134.5\ \mathrm{g/m^2} \end{aligned}$$

The calculation goes on under the climatic conditions of the subsequent months, starting from the hypothesis $(P_{\mathrm{v}} = P_{\mathrm{s}})_{3\mathrm{c}\text{-}4}$ until the total amount of condensate is dried. The results, obtained with the same procedure, are in Table 8.

Step 8 The progressive sum of condensate (Table 8) grows from November to February and then decreases until June, due to evaporation. In the following months, the wall is dry.

Step 9 Assuming the limits for the condensation in mineral wool panels as indicated in the previous paragraph, the maximum level should be

$$\begin{aligned} M_{\mathrm{c.amm}} \leq 5{,}000\ \rho\ \mathrm{L}\ [\lambda/(1 - 1.7\ \lambda)] &= 5{,}000 \times 30 \times 0.03[0.04/(1 - 1.70.04)] \\ &= 193\ \mathrm{g/m^2} \end{aligned}$$

The values in Table 8 show that the sum of the condensate at the end of January would exceed the limit for mineral wool.

Step 10 The assessment is negative, as the condition referred to in Step 8 is satisfied, but not the one in Step 9.

6 Common Typologies of Walls and Their Characteristics: A Wall Database

Considering the method and the results of the projects INVESTIMMO and TABULA, a wall database, which integrates information related both to energy performance (i.e. thermal conductivity, transmittance and dynamic properties) and to hygrothermal behaviour (interstitial condensation assessment) was developed.

The database collects a large number of walls, which represent the constructive typologies of post–World War II buildings all over Europe.

In particular, the database includes information related to:

- thermal parameters: thermal transmittance, thermal mass, areal thermal capacitance;
- dynamic parameters: periodic thermal transmittance, time shift and decrement factor.

The control of moisture problems, such as the interstitial condensation risk, must be analysed basing on specific climatic conditions.

In Table 9 are reported the wall typologies and their parameters. For the calculations, surface thermal resistances of Table 3 are applied.

7 Thermo-Hygrometric Problems: Introduction to Energy Refurbishment

The overall energy consumption of a building is determined by numerous factors, and some of them cannot be changed under renovation. In fact, the geometry, the orientation, the relationship between opaque and transparent surfaces, and the location in urban area represent some of the constraints to the improvement of building energy performance.

In order to reduce energy consumption in existing buildings, the possibilities offered by synergistic actions on elements of the building envelope and plant components have to be assessed.

The HVAC system can be, at least partially, replaced or made more efficient with non-invasive interventions, while the renovation of the envelope can cause more inconveniences to the occupants. Nevertheless, the insulation of the envelope allows to reduce the transmission losses and the energy needs of buildings.

Table 9 Building walls database

01 BM—brick masonry

Layer	L	ρ	λ	R		
1 Internal plaster	2	1,400	0.700	0.028		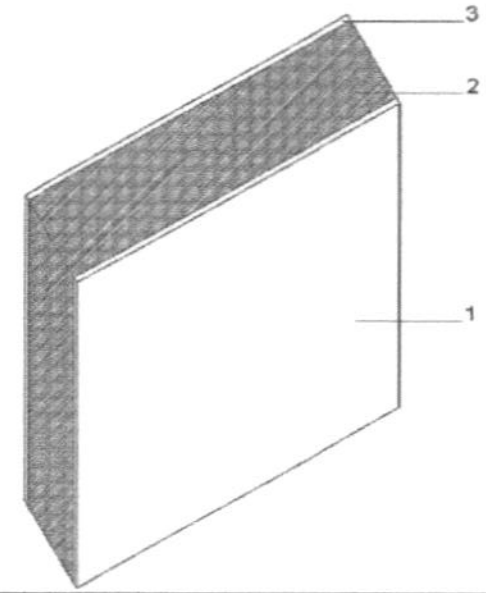
2 Brick	38	1,800	0.720	0.528		
3 External plaster	2	1,800	0.900	0.022		
U	M_s	κ_i	Y_{ie}	S	f_a	
1.336	748	63.1	0.136	14.59	0.102	

02 BM—masonry face brick

Layer	L	ρ	λ	R		
						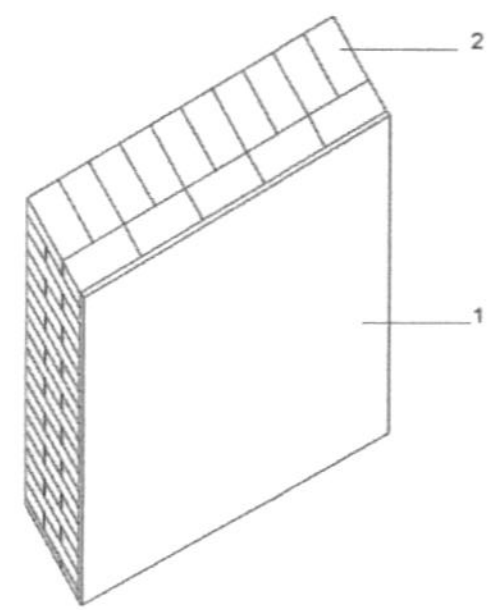
1 Internal plaster	1.5	1,400	0.700	0.021		
2 Brick	51	1,800	0.720	0.708		
U	M_s	κ_i	Y_{ie}	S	f_a	
1.112	939	62.2	0.048	18.5	0.04	

(continued)

Table 9 (continued)

03 BM—cellular brick masonry

Layer	*L*	*ρ*	*λ*	*R*	
1 Internal plaster	2	1,400	0.700	0.028	
2 Brick block	30	1,000	–	0.890[a]	
3 External plaster	2	1,800	0.900	0.022	
U	M_s	κ_i	Y_{ie}	*S*	f_a
0.900	364	53.7	0.197	11.94	0.22

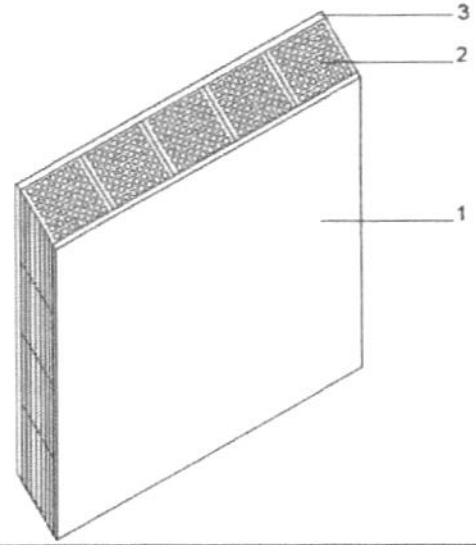

04 SM—stone masonry

Layer	*L*	*ρ*	*λ*	*R*	
1 Internal plaster	2	1,400	0.700	0.028	
2 Stone blocks	80	2,500	2.400	0.333	
3 External plaster	2	1,800	0.900	0.022	
U	M_s	κ_i	Y_{ie}	*S*	f_a
1.805	2,064	72.3	0.033	20.4	0.018

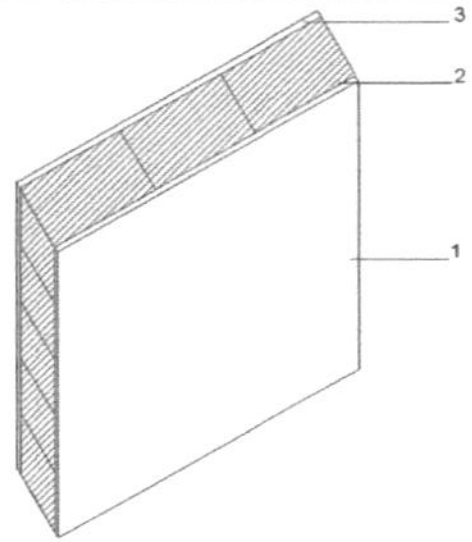

(continued)

Table 9 (continued)

05 SM—tuff stone masonry

Layer	*L*	*ρ*	*λ*	*R*	
1 Internal plaster	2	1,400	0.700	0.028	
2 Tuff stone blocks	50	1,600	0.550	0.909	
3 External plaster	2	1,800	0.900	0.022	
U	M_s	κ_i	Y_{ie}	*S*	f_a
0.885	864	58.7	0.029	20.18	0.033

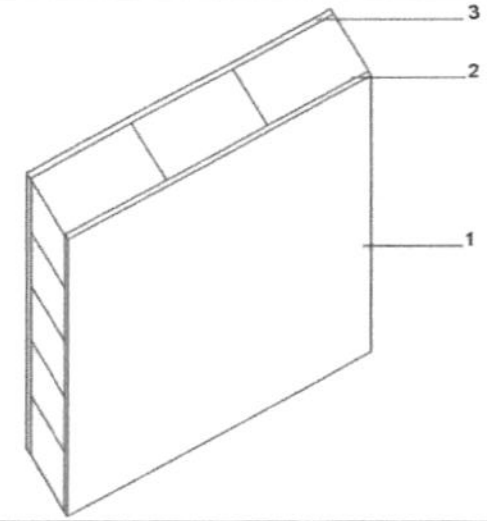

06 SM—stone masonry with air layer

Layer	*L*	*ρ*	*λ*	*R*	
1 Stone masonry	25	2,500	2.400	0.104	
2 Air layer	2.5/30	1.2	–	0.180[b]	
3 Stone masonry	25	2,500	2.400	0.104	
U	M_s	κ_i	Y_{ie}	*S*	f_a
1.791	1,250	84.8	0.121	14.41	0.067

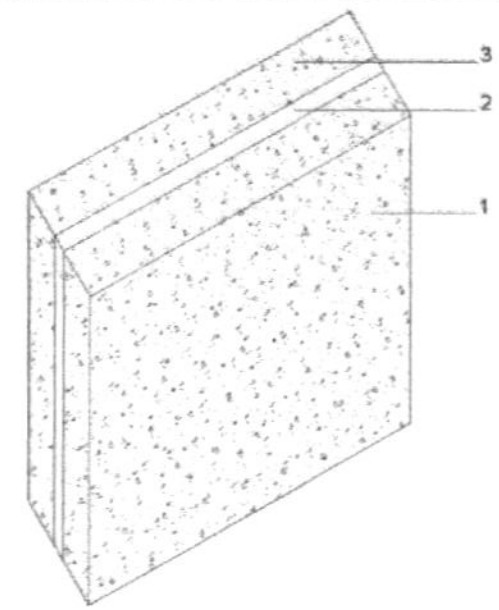

(continued)

Table 9 (continued)

07 SM—stone masonry with insulated layer						
Layer	*L*	*ρ*	*λ*	*R*		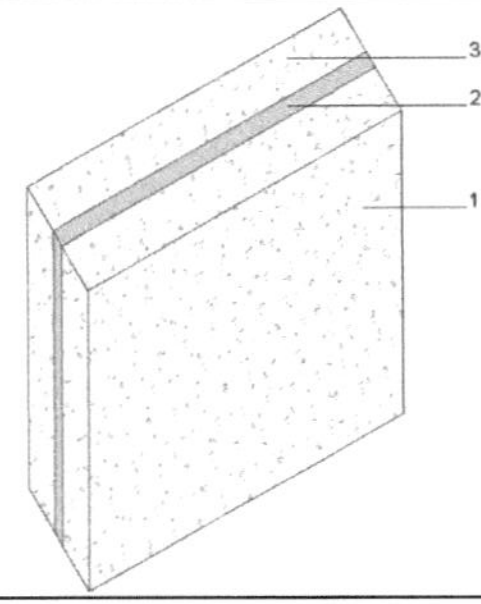
1 Stone masonry	25	2500	2.400	0.104		
2 Insulation	4	30	0.110	0.364		
3 Stone masonry	25	2500	2.400	0.104		
U	M_s	κ_i	Y_{ie}	*S*	f_a	
1.348	1251	84.4	0.070	14.88	0.052	

08 CM—brick and stone masonry						
Layer	*L*	*ρ*	*λ*	*R*		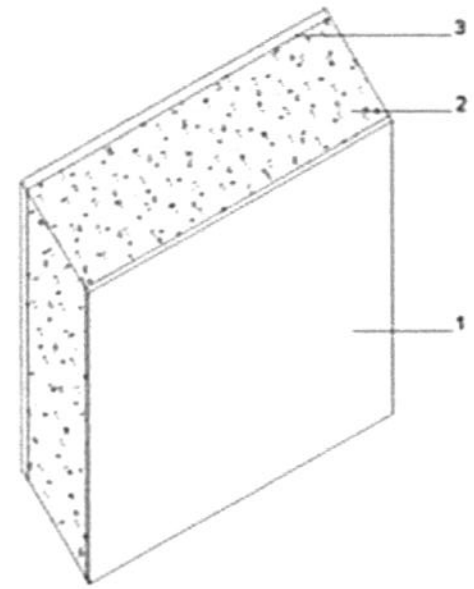
1 Internal plaster	2	1400	0.700	0.028		
2 Brick and stones	50	1500	0.900	0.556		
3 External plaster	2	1800	0.900	0.022		
U	M_s	κ_i	Y_{ie}	*S*	f_a	
1.288	814	62.8	0.104	15.646	0.081	

(continued)

Table 9 (continued)

09 CM—masonry with air layer					
Layer	*L*	*ρ*	*λ*	*R*	
1 Internal plaster	2	1,400	0.700	0.028	
2 Hollow bricks	8	800	–	0.200[a]	
3 Filler	10	1,500	0.700	0.143	
4 Bricks	25	1,800	0.720	0.347	
5 External plaster	2	1,800	0.900	0.022	
U	M_s	κ_i	Y_{ie}	*S*	f_a
1.098	728	49.4	0.093	15.53	0.084

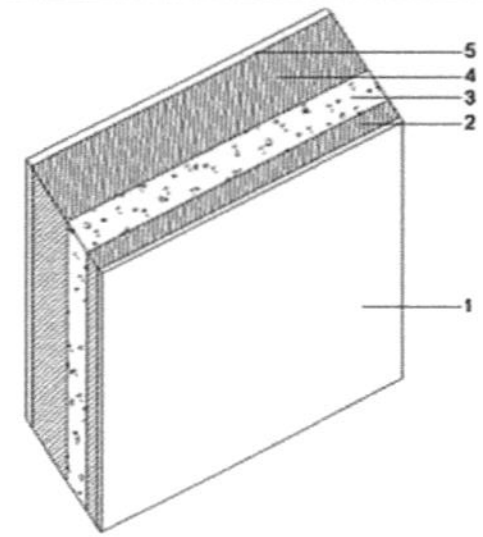

10 CM—hollow concrete block masonry (1)					
Layer	*L*	*ρ*	*λ*	*R*	
1 Internal plaster	2	1,400	0.700	0.028	
2 Hollow concrete blocks	20	1,400	0.500	0.40	
3 External plaster	2	1,800	0.900	0.022	
U	M_s	κ_i	Y_{ie}	*S*	f_a
1.611	344	65.1	0.656	8.12	0.407

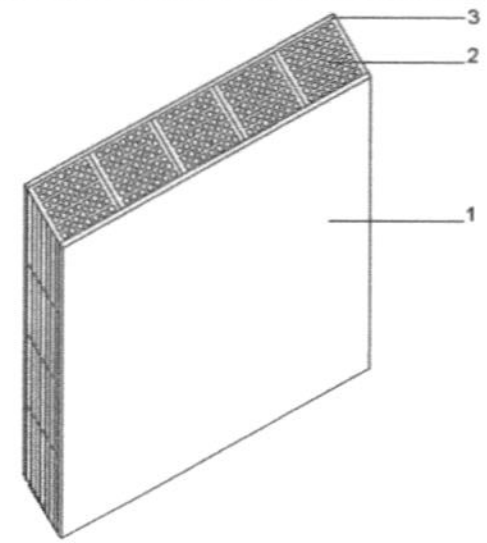

(continued)

Table 9 (continued)

11 CM—hollow concrete block masonry (2)					
Layer	*L*	*ρ*	*λ*	*R*	
1 Internal plaster	2	1,400	0.700	0.029	
2 Hollow bricks	8	800	–	0.200[a]	
3 Air layer	2.5/30	1.2	–	0.180[b]	
4 Concrete blocks	20	1,400	0.500	0.40	
U	M_s	κ_i	Y_{ie}	*S*	f_a
1.022	372	53.3	0.258	10.8	0.252

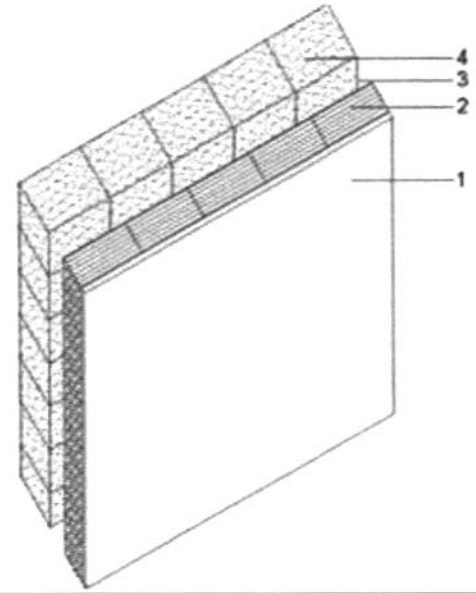

12 HB—hollow brick masonry (1)					
Layer	*L*	*ρ*	*λ*	*R*	
1 Internal plaster	2	1,400	0.700	0.028	
2 Hollow bricks	8	800	–	0.200[a]	
3 Air layer	2.5/30	1.2	–	0.180[b]	
4 Hollow bricks	12	800	–	0.310[a]	
5 External plaster	2	1,800	0.900	0.022	
U	M_s	κ_i	Y_{ie}	*S*	f_a
1.098	224	57.9	0.594	7.268	0.541

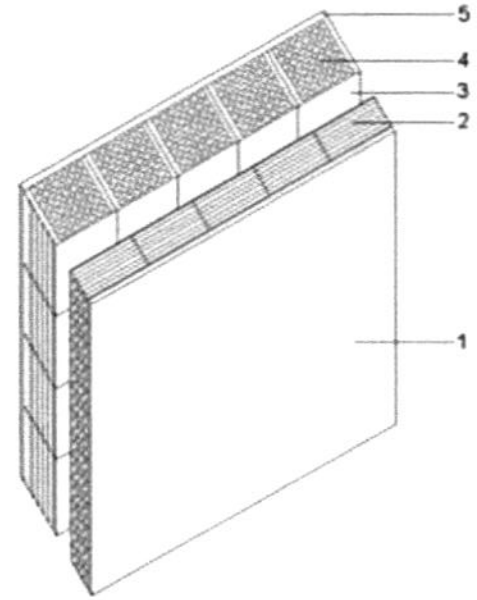

(continued)

Table 9 (continued)

13 HB—air layer brick masonry (1)						
Layer	*L*	ρ	λ	*R*		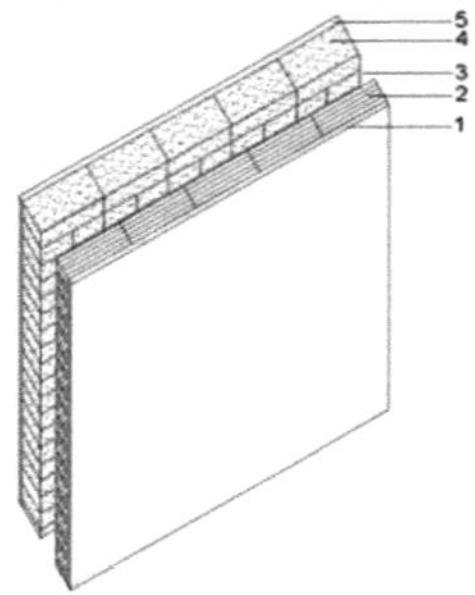
1 Internal plaster	2	1,400	0.700	0.028		
2 Hollow bricks	8	800	–	0.200[a]		
3 Air layer	2.5/30	1.2	–	0.180[b]		
4 Bricks	25	1,800	0.720	0.347		
5 External plaster	2	1,800	0.900	0.022		
U	M_s	κ_i	Y_{ie}	*S*	f_a	
1.055	578	50.6	0.152	13.12	0.144	

14 HB—air layer brick masonry (2)						
Layer	*L*	ρ	λ	*R*		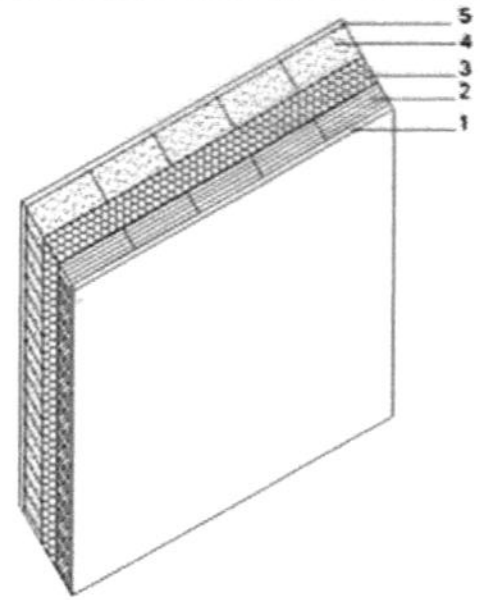
1 Internal plaster	2	1,400	0.700	0.028		
2 Hollow bricks	8	800	–	0.200[a]		
3 Polystyrene	4	15	0.045	0.889		
4 Bricks	25	1,800	0.720	0.347		
5 External plaster	2	1,800	0.900	0.022		
U	M_s	κ_i	Y_{ie}	*S*	f_a	
0.604	578	53.5	0.063	14.22	0.105	

(continued)

Table 9 (continued)

15 HB—air layer brick masonry (3)

Layer	*L*	*ρ*	*λ*	*R*	
1 Internal plaster	2	1,400	0.700	0.028	
2 Hollow bricks	12	800	–	0.310[a]	
3 Air layer	2.5/30	1.2	–	0.180[b]	
4 Hollow bricks	12	1,200	–	0.310[a]	
U	M_s	κ_i	Y_{ie}	*S*	f_a
1.001	268	56.7	0.379	9.15	0.378

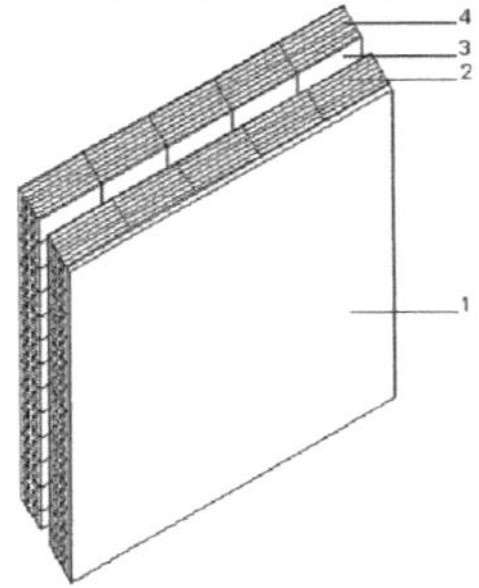

16 HB—air layer brick masonry (4)

Layer	*L*	*ρ*	*λ*	*R*	
1 Internal plaster	2	1,400	0.700	0.028	
2 Hollow bricks	12	800	–	0.310[a]	
3 Air layer	2.5/30	1.2	–	0.180[b]	
4 Bricks	12	1,800	0.720	0.720	
U	M_s	κ_i	Y_{ie}	*S*	f_a
1.169	340	57.0	0.45	9.05	0.385

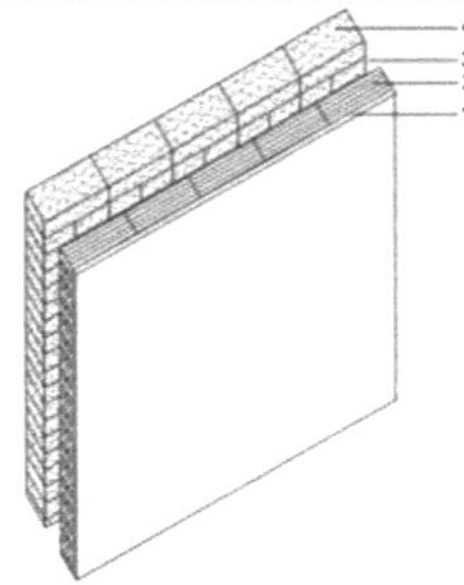

(continued)

Table 9 (continued)

17 PC—concrete wall						
Layer	**L**	ρ	λ	**R**		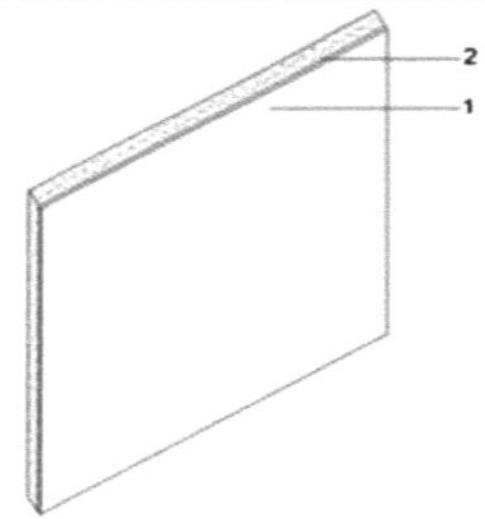
1 Internal plaster	1	1,400	0.700	0.028		
2 Concrete wall	25	1,400	0.720	0.528		
U	M_s	κ_i	Y_{ie}	S	f_a	
1.837	364	64.7	0.599	8.46	0.367	

(continued)

Table 9 (continued)

18 PC—concrete insulated wall (2)						
Layer	*L*	*ρ*	*λ*	*R*		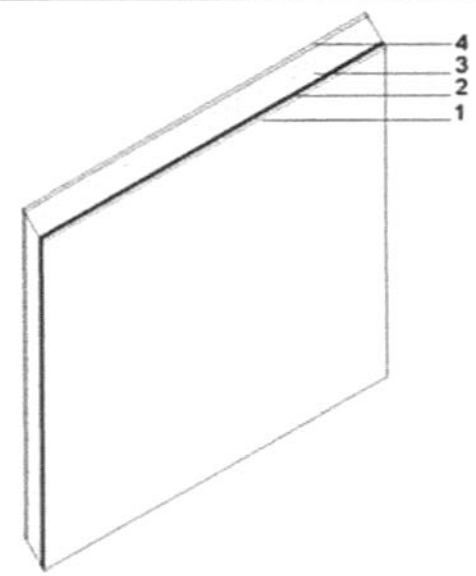
1 Internal plaster	1	1,400	0.700	0.014		
2 Mineral wall panel	2	30	0.040	0.50		
3 Concrete panel	25	1,400	0.580	0.431		
4 External plaster	2	1,800	0.900	0.022		
U	M_s	$κ_i$	Y_{ie}	*S*	f_a	
0.879	400	25.6	0.181	10.25	0.206	

L Thickness (cm)
ρ Density (kg/m^3)
λ Thermal conductivity [W/(m K)]
R Thermal resistance (m^2 K/W)
U Thermal transmittance [W/(m^2 K)]
M_s Specific mass (kg/m^2)
$κ_i$ Arei thermal capacity. [kJ/(m^2 K)]
Y_{ie} Dynamic thermal transmittance [W/(m^2 K)]
S Time shift (h)
f_a Decrement factor (-)
[a] Estimated values of thermal equivalent resistance for hollow bricks
[b] Values for unventilated air layers from the Standard EN ISO 6946

The building energy performance assessment requires also the analysis of HVAC systems, in order to verify whether the plants can efficiently operate even with a low energy demand. On the other hand, reducing the amount of energy allows to consider alternative system solutions that include the exploitation of renewable sources. The refurbishment of the opaque elements of existing buildings means, in general, using thermal insulation materials to decrease the transmittance and therefore the energy losses through the walls facing towards the external environment. The consequences of the insulation can be summarized with the following considerations.

Increasing insulation of external structures allows the reduction of winter heat losses and of summer loads, with the reduction of global energy consumption.

Moreover, a careful design and verification of glazed surfaces and their solar shading is necessary. In fact, incorrect evaluation of the transparent elements influence in summer could cause overheating and greenhouse effects in the indoor environment. In this case, a high thermal insulation level would reduce the heat loss through the walls from inside to outside, maintaining a high temperature, not desired, in the indoor environments, in summer.

Insulating the external envelope can improve the indoor comfort as if the wall surface temperature is closer to the air one, the heat exchange by radiation between the human body and the different surfaces is reduced.

There are three main solutions for thermal insulation:

- external insulation;
- internal insulation;
- air layer insulation.

The most effective one is the external insulation, even if it is also the most expensive. Moreover, it is not always feasible, since, for the application, the facade must be free of ornaments. This solution is interesting if the installation cost of scaffolding and other works is considered in association with the other renovation actions already planned and if there is not a large number of overhang elements in the external surface.

With the installation of a ventilated facade, good results can be reached, even if with rather high costs. The intervention with insulating plaster could be useful, if considered synergistically with other actions, due to the low thickness of the layer. In fact, even if insulating plaster is characterized by a low thermal conductivity, it could lead to results that do not comply the minimum requirements for the envelope.

The air layer insulation can be interesting for many buildings built between the 1960s and 1970s, since they often have empty cavity in the envelope layers, which can be filled by melted materials. The results can be effective: it is possible to apply the insulation from external holes in the façade or from inside without high costs.

However, the thermal properties of the insulation (foam or melted material) have to be defined. Moreover, before performing this kind of action, it would be appropriate to have information on the effective period in which the properties are guaranteed; otherwise, it will be necessary a second application after a period in

Table 10 Hygrothermal properties of insulation materials

Type	λ [W/(m K)]	δ [kg/(m s Pa)]	ρ (kg/m^3)
Internal or external insulation			
Mineral wall (MW)	0.036	193×10^{-12}	40
Wooden fibreboard (WF)	0.040	97×10^{-12}	110
Polystyrene (PY)	0.033	1.3×10^{-12}	35
Air layer insulation			
Cellulose	0.038	1×10^{-12}	35
Polyurethane foam	0.030	3.8×10^{-12}	30

dependence of the degeneration and the maintained uniformity of the filling material.

The internal insulation may be less difficult, since it regards single dwellings and it does not involve actions on the facade, the use of scaffolding, etc. Nevertheless, this intervention could reduce the internal surface of the living spaces; therefore, it has to be correctly designed and planned. In fact, while the insulation in the cavity presents physical limits due to the width of the air layer, the application of insulating panels becomes effective for thicknesses generally higher than 4–6 cm, especially in the continental climatic conditions.

Regarding the dynamic thermal performance of the walls, the position of the insulation greatly influences the thermal inertia of the structure. The heat storage is determined by the properties of materials, which are involved in the heating transmission of the envelope, In general, the external insulation provides better results.

As regards hygrometric properties, a higher thermal insulation undoubtedly involves a good improvement in terms of surface moisture problems. Nevertheless, the situation should be carefully assessed in terms of risk of interstitial condensation, especially when the insulating layer is located close to the internal side of the wall.

7.1 Evaluation of Insulation Intervention: Practical Examples

As reference for the thermal transmittance reductions of building walls, the effects of the thermal insulation of the structures of the database (Table 9) are calculated, considering two kind of installation, on the internal or the external side. As an alternative, for air layer masonry, the effect of the cavity insulation is also considered.

For each structure, three different kinds of insulated materials widely used in building refurbishment are considered (Table 10). Their properties are gathered from the technical files of commercial products.

Moreover, for the evaluation of the periodic thermal transmittance, it is also taken into account the position of the insulating layer (external or internal). An additional layer of external plaster is added in case of external insulation.

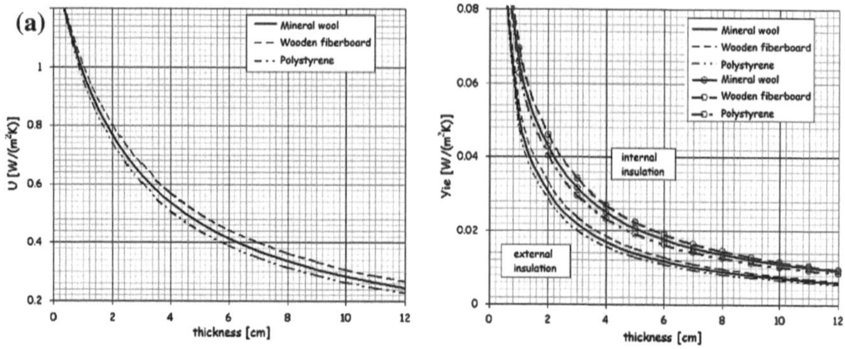

Fig. 7 U and Y_{ie} values versus insulation thickness. **a** 01 BM—brick masonry. **b** 02 BM—masonry face brick. **c** 03 BM—cellular brick masonry. **d** 04 SM—stone masonry. **e** 05 SM—tuff stone masonry. **f** 06 SM—stone masonry with air layer. **g** 06 SM—stone masonry with air layer—cavity insulation. **h** 07 SM—stone masonry with insulated layer. **i** 08 CM—brick and stone masonry. **j** 09 CM—masonry with air layer. **k** 10 CM—hollow concrete block masonry (1). **l** 11 CM—hollow concrete block masonry (2). **m** 11 CM—hollow concrete block masonry (2)—cavity insulation. **n** 12 HB—hollow brick masonry (1). **o** 12 HB—hollow brick masonry (1)—cavity insulation. **p** 13 HB—air layer brick masonry (1). **q** 13 HB—air layer brick masonry (1)—cavity insulation. **r** 14 HB—air layer brick masonry (2). **s** 15 HB—air layer brick masonry (3). **t** 15 HB—air layer brick masonry (3)—cavity insulation. **u** 16 HB—air layer brick masonry (4). **v** 16 HB—air layer brick masonry (4)—cavity insulation. **w** 17 PC—concrete wall. **x** 18 PC—concrete insulated wall (2)

The following graphs (Fig. 7a–x) allow to check the insulation thickness needed to improve the performances indicated by local requirements or limits of the thermal transmittance U and periodic thermal transmittance Y_{ie}. In addition, cavity filling is evaluated, where applicable.

8 Interstitial Condensation Evaluation: Maximum Insulating Thickness Allowed in the Refurbishment for Some Wall Structures

It is important to associate the choice of insulation, in relation to the limits for the energy refurbishment, to the corresponding hygrometric assessment. Sometime, for particular climatic conditions and building structures, an optimal choice in terms of energy saving may not be suitable in relation to the risk of interstitial condensation.

Following the steps of the procedure for the interstitial condensation assessment, the previously described walls are considered in the following calculations, to determine the insulating layer thicknesses for a positive assessment (Step 10). As reference, only the structure number is indicated, from 01 to 18.

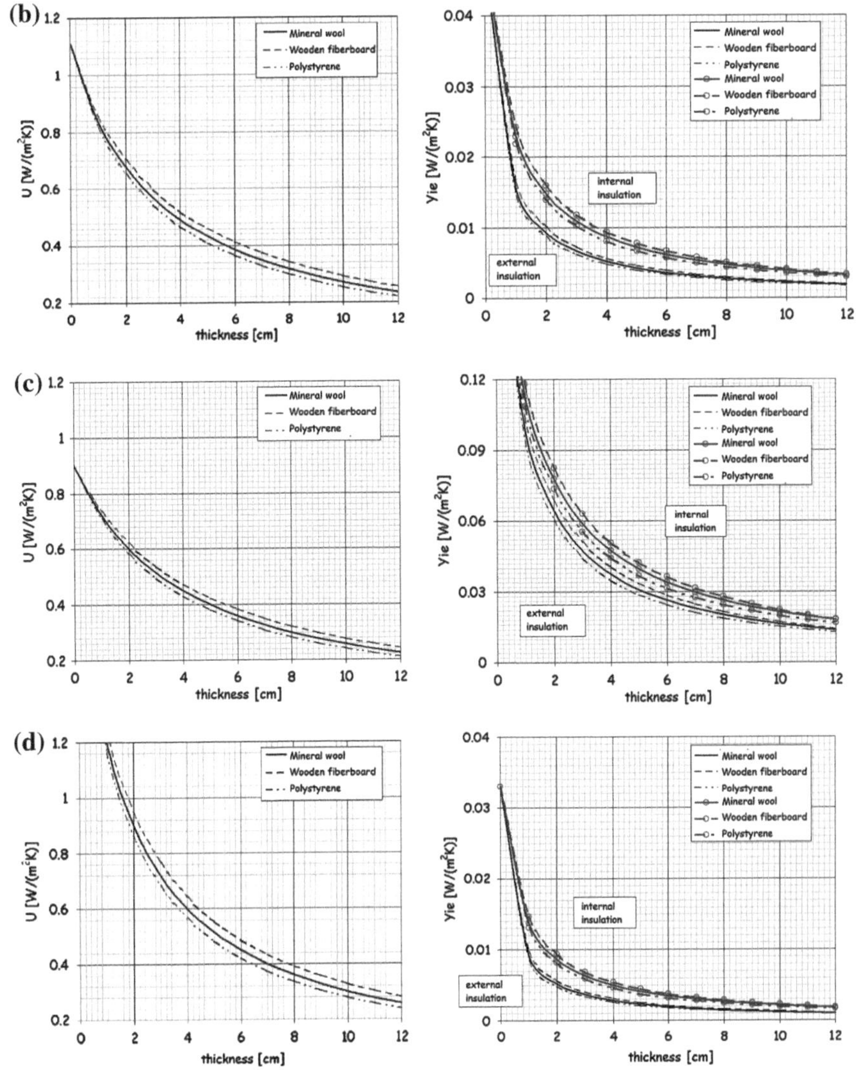

Fig. 7 continued

As before, for the thermal calculations, various insulating materials, thicknesses (from 1 to 12 cm), position (outside, inside, filling the air layer) have been considered with their hygrothermal parameters.

The indoor moisture production was referred to humidity class 3 (buildings with unknown occupancy, Fig. 4). The climatic data of six different regions in Europe are resumed in Table 11 [t_e in (°C), P_{ve} in (Pa)].

The upper limits of condensation in the annual cycle for different materials have been calculated in function of thickness L (m), density ρ (kg/m^3) and conductivity

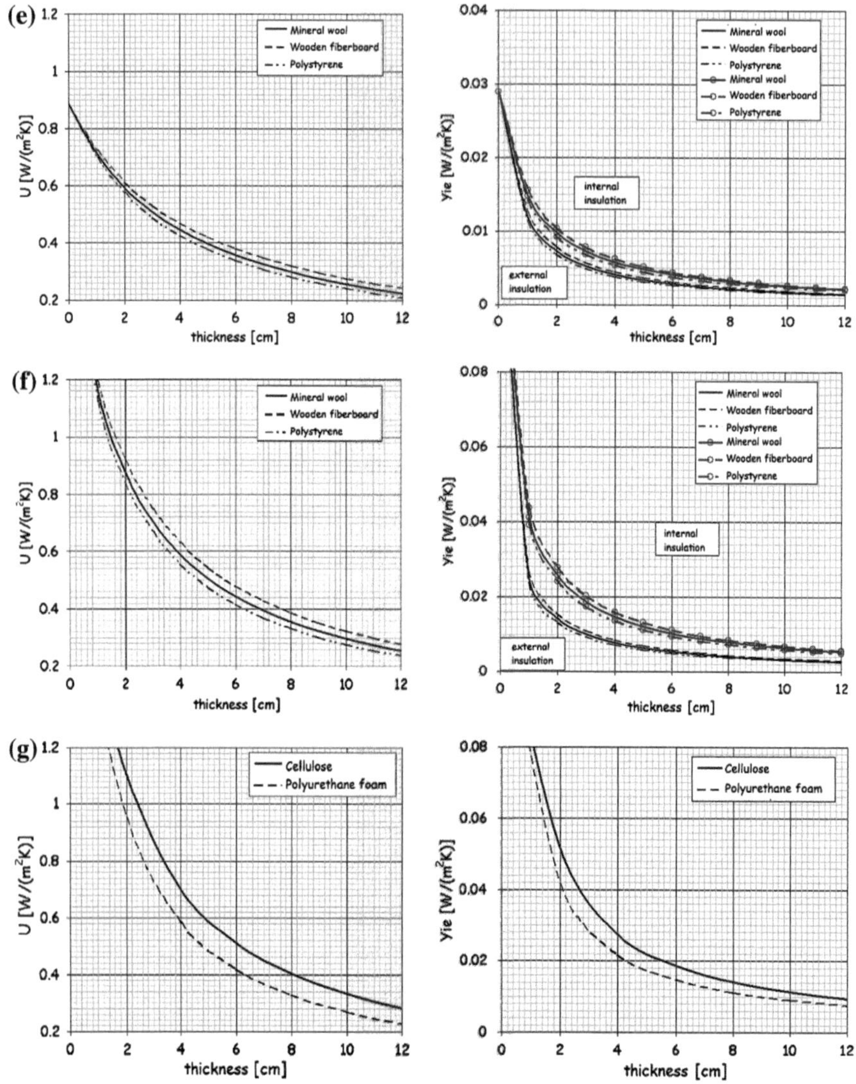

Fig. 7 continued

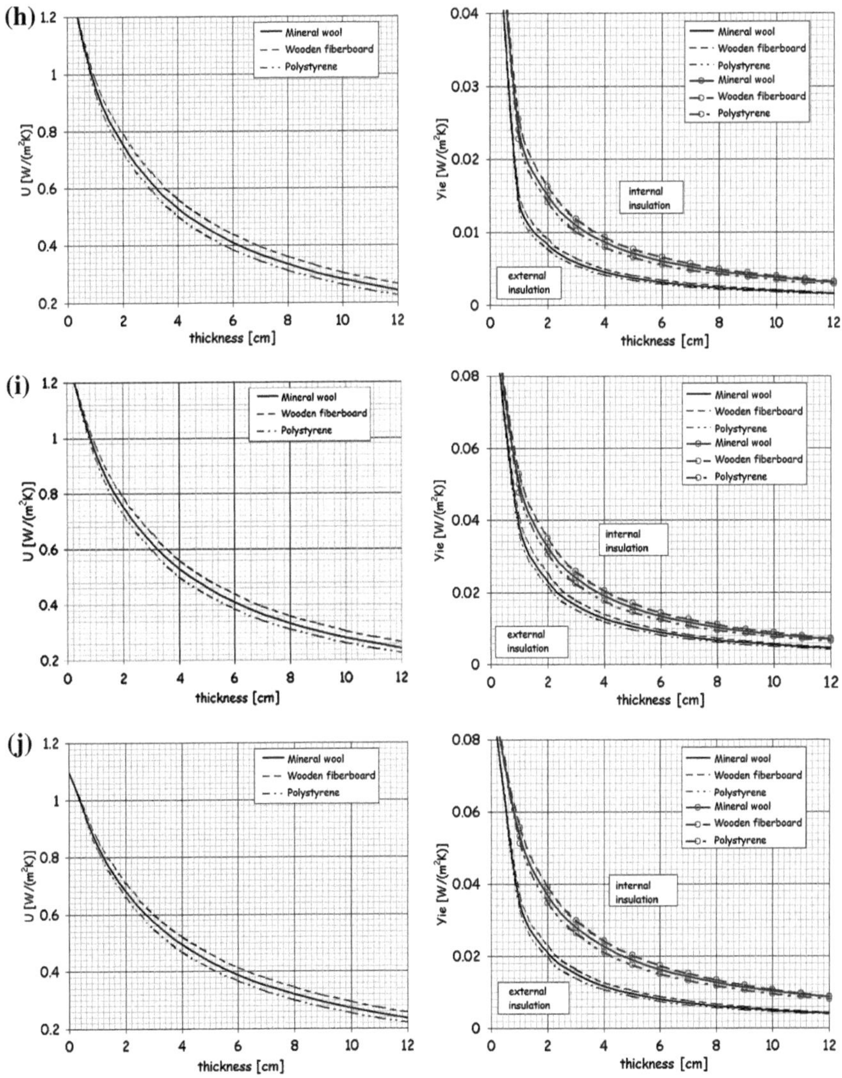

Fig. 7 continued

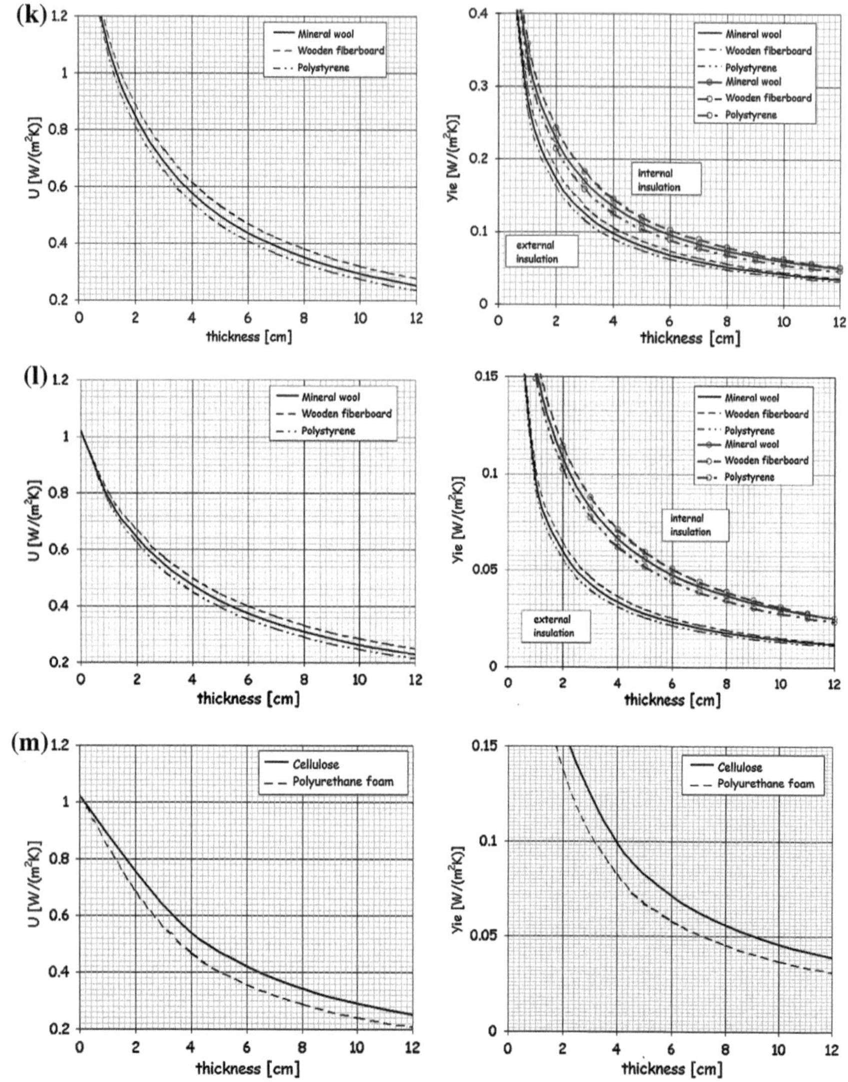

Fig. 7 continued

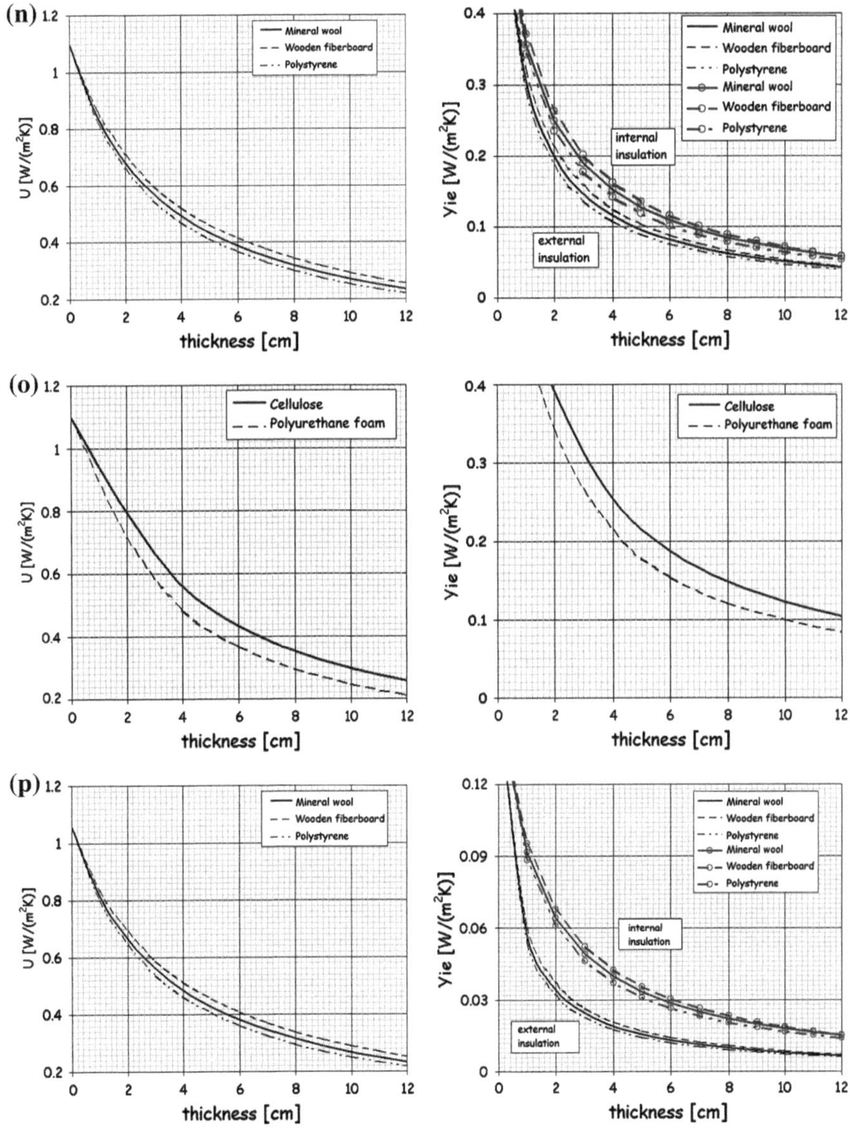

Fig. 7 continued

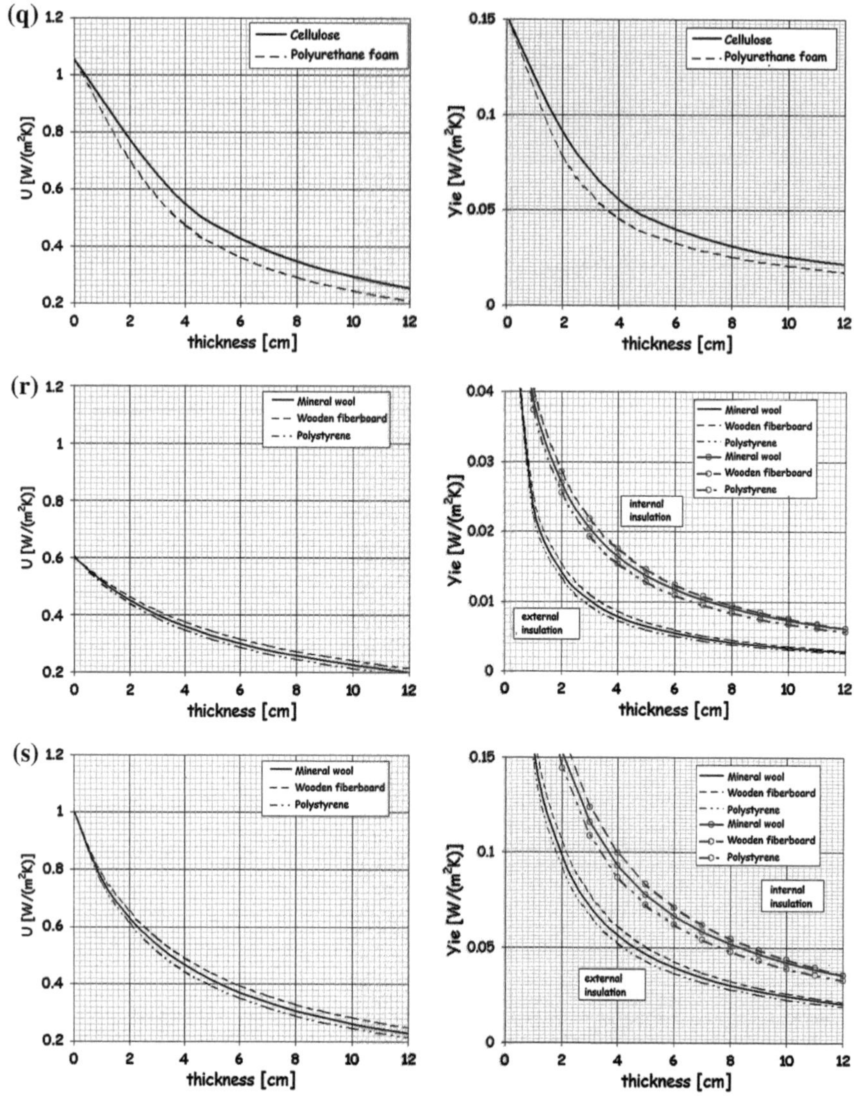

Fig. 7 continued

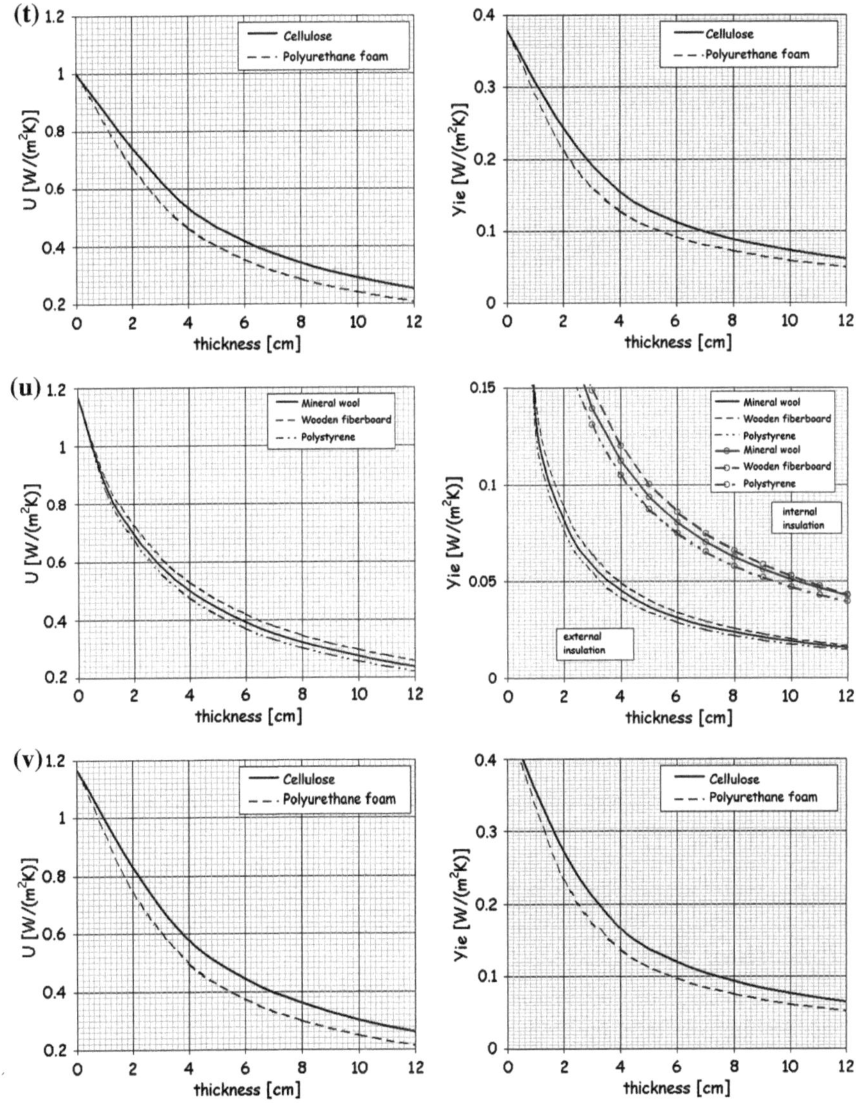

Fig. 7 continued

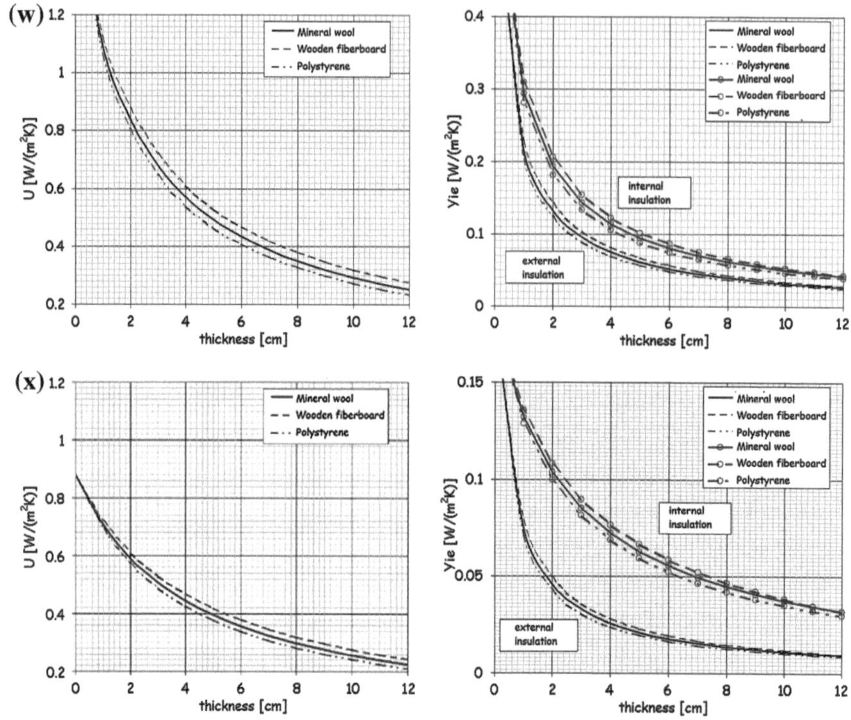

Fig. 7 continued

λ [W/(m K)] of the chosen materials, in agreement with the indications described in Sect. 5 (Table 12).

8.1 Existing Walls

Condensation occurs in the following existing structures:

- Wall 06, in the climatic conditions of Milan and Hamburg (<2 g/m^2).
- Wall 07, in the climatic conditions of Milan, Paris, Brussels (<3 g/m^2) and Hamburg (<6 g/m^2).
- Wall 11 and 18, in the climatic conditions of Milan (<110 and <350 g/m^2, respectively), Hamburg (<200 and <2,000 g/m^2), Paris and Brussels (<35 and <720 g/m^2).
- Wall 14 and 16, in the climatic conditions of Milan (<20 and <4 g/m^2, respectively).

Table 11 Climatic conditions considered in the calculations

Climate		Jan	Feb	Mar	Apr	May	Jun	Jul	Aug	Sep	Oct	Nov	Dec
Rome	t_e	8.3	9.0	10.6	13.3	17.6	21.2	24.2	24.4	21.5	17.0	12.8	9.4
	P_{ve}	833.0	932.7	1042.1	1167.5	1523.9	1915.8	2354.7	2263.8	1962.0	1500.3	1153.5	953.3
Milan	t_e	0.3	2.2	7.5	10.9	16.3	19.7	23.1	22.2	17.6	12.5	6.5	2.2
	P_{ve}	556.8	479.4	698.1	885.2	1285.3	1857.4	1820.5	1817.2	1609.6	1300.7	803.6	523.3
Paris	t_e	3.9	4.2	7.0	10.0	14.3	16.8	19.4	19.7	15.7	11.3	6.4	4.5
	P_{ve}	735.4	558.7	754.7	891.2	1081.4	1486.4	1469.5	1555.8	1387.2	1090.8	859.3	754.3
Madrid	t_e	5.6	6.9	10.0	11.7	16.9	20.6	25.5	24.7	20.1	14.2	9.4	5.9
	P_{ve}	656.5	690.5	744.7	762.5	1151.7	1257.8	1293.8	1407.5	1339.6	1133.4	884.9	773.6
Brussels	t_e	3.1	3.2	6.4	8.9	12.9	15.6	18.4	17.4	14.5	10.9	6.6	4.9
	P_{ve}	644.6	663.8	804.8	868.6	1141.3	1320.4	1640.9	1492.9	1333.7	1047.6	817.6	754.7
Hamburg	t_e	2.0	0.2	4.4	6.9	12.8	15.8	16.9	17.4	13.5	9.8	5.2	2.6
	P_{ve}	630.9	506.3	682.6	737.9	1042.3	1328.1	1393.2	1511.5	1255.4	988.0	726.8	658.1

Table 12 Vapour condensation upper limits for some building materials

Thickness		L (cm) =	1	2	3	4	5	6	7	8	9	10	11	12
	ρ (kg/m^3)	λ [W/(mK)]	Q_{amm} (g/m^2)											
Brick			500	500	500	500	500	500	500	500	500	500	500	500
Concrete			500	500	500	500	500	500	500	500	500	500	500	500
Timber	100	0.040	30	60	90	120	150	180	210	240	270	300	330	360
Plaster and mortar	1,400	0.700	420	500	500	500	500	500	500	500	500	500	500	500
Organic fibre														
water-resistant adhesive	100	0.040	20	40	60	80	100	120	140	160	180	200	220	240
not water-resistant adhesive	100	0.040	5	10	15	20	25	30	35	40	45	50	55	60
Mineral fibre	40	0.036	77	153	230	307	383	460	500	500	500	500	500	500
Cellular plastic materials	35	0.033	61	122	184	245	306	367	428	489	500	500	500	500

8.2 External Insulation

The results of the assessment of condensation in the case of external insulation show that sometime condensation may occur before the external plaster layer with moderate quantities.

Walls 1–2, 4–11 and 13: no condensation occurs for any climatic conditions.

Wall 3:1 cm PY produces condensation in the climatic conditions of Milan (<30 g/m^2) and Hamburg (<15 g/m^2).

Wall 12 presents condensation in the climatic conditions of Milan, for MW and WF (thickness >2 cm), and of Hamburg for MW (>5 cm). In the same climatic conditions, 1 cm PY produces, respectively, <35 and <15 g/m^2 of condensation in the annual cycle. In any case, the amount of condensation is moderate and lower than the maximum values for the materials involved.

Wall 14, affected by interstitial condensation in the existing conditions and climate of Milan, becomes free of condensation for the same climate, if MW or WF is used. For PY, the amount of assessed condensation reduces gradually to zero, but it occurs for 1–2 cm.

Walls 15, 16 and 17 present condensation for MW (for thickness >4, >5 and >4 cm, respectively), for WF (>7, >8, >5 cm) and for PY (1 cm only for 15 and 16) in the climatic conditions of Milan. Wall 15 presents condensation also for Hamburg only for 1 cm thickness PY (<20 g/m^2). In any case, the amount of condensation is small and lower than the maximum values for the materials involved.

Wall 18: the problem of condensation, found for Paris and Brussels in the existing conditions, disappears with any insulating material and thickness, while for Milan remains also with 1 cm of MW, WF and PY. For the climatic conditions of Hamburg, condensation occurs only for 1 cm PY.

8.3 Internal Insulation

Generally, significant problems of condensation emerge for internal insulation, even if sometimes it is an easier refurbishment solution, mostly in the case of articulated, painted, decorated façades. The calculation results are summarized in Table 13: for each wall (01–18) and the three insulating materials, the thickness that corresponds to the absence of condensation is in brackets, the positive assessment is pointed out as X, on the contrary the number indicates the maximum thickness that allows a positive assessment.

Table 13 Maximum thickness (cm) of insulating layer that satisfies the two criteria for interstitial condensation (acceptable maximum condensate quantity and complete evaporation)-Internal insulation

		Rome	Milan	Paris	Madrid	Brussels	Hamburg
01	MW	X(6)	1(1)	2(1)	7(3)	2(2)	1(1)
	WF	X(7)	1(1)	2(2)	4(3)	2(2)	1(1)
	PY	X	X	X	X	X	X
02	MW	X(7)	2(1)	2(2)	7(3)	2(2)	2(1)
	WF	X(9)	2(1)	2(2)	5(4)	3(2)	2(2)
	PY	X	X	X	X	X	X
03	MW	X(11)	4(1)	5(3)	X(5)	5(3)	3(2)
	WF	X	3(2)	4(3)	X(6)	5(4)	3(3)
	PY	X	X	X	X	X	X
04	MW[a]	4(2)	0(0)	0(0)	2(1)	0(0)	0(0)
	WF[a]	3(3)	0(0)	0(0)	1(1)	1(0)	0(0)
	PY[a]	X(2)	X(0)	X(0)	X(1)	X(0)	1–2 only(0)
05	MW	X(8)	3(1)	4(3)	10(4)	4(3)	2(2)
	WF	X(10)	2(2)	3(3)	8(5)	3(3)	3(2)
	PY	X	X	X	X	X	X
06	MW[a]	5(3)	0	1(0)	2(1)	1(0)	0(0)
	WF[a]	4(3)	0	1(0)	1(1)	1(0)	0(0)
	PY[a]	X(3)	X(0)	X(0)	X(1)	X(0)	2 only(0)
07	MW[a]	9(4)	0(0)	1(0)	4(0)	1(0)	1(0)
	WF[a]	7(5)	0(0)	1(0)	3(0)	1(0)	1(0)
	PY[a]	X(5)	X(0)	X(0)	X(0)	X(0)	X(0)
08	MW	X(6)	1(1)	2(1)	7(3)	2(2)	1(1)
	WF	X(7)	1(1)	2(2)	4(3)	2(2)	1(1)
	PY	X	X	X	X	X	X
09	MW	X(8)	3(0)	3(1)	10(3)	3(2)	1(1)

(continued)

Table 13 (continued)

		Rome	Milan	Paris	Madrid	Brussels	Hamburg
	WF	X(9)	1(1)	2(1)	7(4)	2(2)	2(1)
	PY	X	X	X	X	X	X
10	MW	6(3)	0	1	2(1)	1(1)	1
	WF	5(4)	0	1	2(2)	1(1)	1
	PY	X	X(1)	X(1)	X	X(2)	X(3)
11	MW	X(2)	0(0)	0(0)	X(0)	0(0)	0(0)
	WF	6(3)	0(0)	0(0)	1(0)	0(0)	0(0)
	PY	X	X(0)	X(0)	X	X(0)	1 only(0)
12	MW	X(10)	2(0)	3(1)	X(4)	3(2)	1(1)
	WF	X	1(0)	2(1)	X (5)	3(2)	1(1)
	PY	X	X	X	X	X	X
13	MW	X(8)	0(0)	1(0)	X(2)	1(1)	0(0)
	WF	X(11)	0(0)	1(0)	8(2)	1(1)	0(0)
	PY	X	X	X	X	X	X
14	MW[a]	X	8(0)	8(0)	X(6)	8(1)	6(0)
	WF[a]	X	0(0)	8(0)	X(8)	8(1)	0(0)
	PY[a]	X	X	X	X	X	X
15	MW	X(11)	3(1)	4(2)	X(5)	4(3)	3(2)
	WF	X	3(1)	4(3)	X(6)	4(4)	3(2)
	PY	X	X	X	X	X	X
16	MW	X(9)	2 only(0)	3(0)	X(4)	3(1)	0(0)
	WF	X	0(0)	1(0)	X(5)	2(1)	0(0)
	PY	X	X	X	X	X	X
17	MW	6(3)	0(0)	1(1)	2(1)	1(1)	0(0)
	WF	6(4)	0(0)	1(1)	2(2)	1(1)	0(0)
	PY	X	X	X	X	X	X

(continued)

Table 13 (continued)

		Rome	Milan	Paris	Madrid	Brussels	Hamburg
18	MW	3(2)	0(0)	0(0)	0(0)	0(0)	0(0)
	WF	3(2)	0(0)	0(0)	0(0)	0(0)	0(0)
	PY	X	X > 2	X > 2	X(0)	X(0)	X(0)

X maximum limit verified for thickness up to 12 cm (no condensation at all), *MW* mineral wool, *WF* wooden fibreboard, *PY* polystyrene panels
In brackets, maximum thickness (cm) without any condensation all over the year
only condensation occurs only in the case of one or two thicknesses and not for other ones
[a] condensation occurs also on the external face of the existing insulating layer

Table 14 Maximum amount of interstitial condensation (g/m^2), and in brackets, the corresponding air layer thickness (cm)

		Rome	Milan	Paris	Madrid	Brussels	Hamburg
06	Cell	0.0	6.7(7–8)	7.4(7–9)	1.0(7–12)	8.3(8)	12.6(7)
	Foam	0.0	11.3(12–14)	11.2(11–15)	2.2(11–20)	13.7(11–15)	18.5(10–14)
11	Cell	0.0	15.6(3)	7.3(3)	0.0	3.6(3)	27.5(3)
	Foam	0.0	189.3(3)	164.7(3)	34.9(5)	200.2(3)	298.7(3)

8.4 Air Cavity Insulation

In the case of cavity insulation, the risk of condensation is estimated for 3–30 cm air layer thickness and for cellulose or polyurethane foam filling. Through the calculations, 12, 13 and 15 walls are free of condensation for all the climatic conditions considered (Table 11), even if 13 wall presents condensation for 3–11 cm cavity thickness with PY foam, in the climatic conditions of Milan. Condensation results in 06 and 11 walls are resumed in Table 14.

8.5 Refurbishment Example: How to Evaluate the Intervention

Considering the wall described in Table 15 (taken from Table 9), it is shown how to calculate the appropriate thickness of mineral wool, to reach the value of thermal transmittance $U = 0.30$ W/(m^2 K) and the dynamic thermal transmittance $Y_{ie} = 0.12$ W/(m^2 K). Interstitial condensation assessment is also performed.

Starting from the thermal properties of the analysed wall, the minimum thickness of insulation which allows to reach the imposed transmittance values is determined through the diagrams in Fig. 8a and b.

To reach the U-value of 0.30 W/(m^2 K), 7 cm insulation thickness is necessary both for the internal and the external insulation layer.

For the dynamic thermal transmittance, 4 cm thickness has to be applied for the external layer and 5.5–6 cm is needed if the intervention is on the internal layer.

This last information is useful to define the optimal position of the insulation layer (internal or external), even if in this case the most strict condition is determined by the thermal transmittance (Table 16).

Moreover, an intervention of air layer insulation is also evaluated (Fig. 9). If the air cavity thickness is greater than 10 cm, the target U-value is reached both with cellulose or polyurethane foam filling. The same result is obtained for the dynamic thermal transmittance limit.

Referring to the climatic conditions in Table 11, the application of 7 cm mineral wool insulating layer causes interstitial condensation. For the external insulation, it occurs in the climatic conditions of Milan and of Hamburg. In any

Table 15 Thermal properties of the wall

12 HB—Hollow brick masonry (1)					
Layer	*L (cm)*	*ρ (kg/m³)*	*λ (W/m K)*	*R (m² K/W)*	
1 Internal plaster	2	1,400	0.700	0.028	
2 Hollow bricks	8	800	–	0.200[a]	
3 Air layer	2.5/30	1.2	–	0.180[b]	
4 Hollow bricks	12	800	–	0.310[a]	
5 External plaster	2	1800	0.900	0.022	
U[W/(m² K)]	*M_s (kg/m²)*	*$κ_i$ [kJ/(m² K)]*	*Y_{ie} [W/(m² K)]*	*S (h)*	*f_a (-)*
1.098	224	57.9	0.594	7.268	0.541

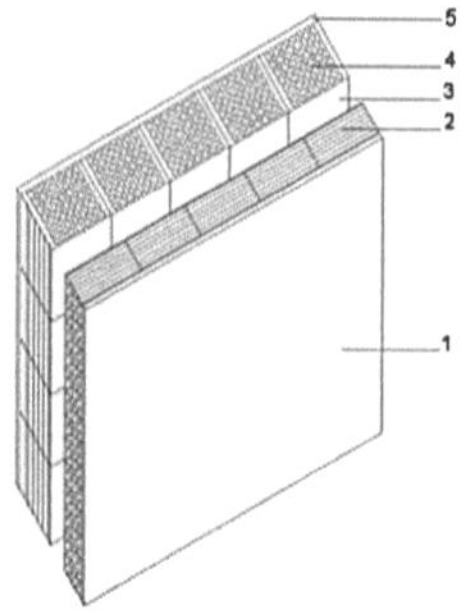

[a] Equivalent thermal resistance for hollow bricks
[b] Thermal resistance for unventilated air layer EN ISO 6946

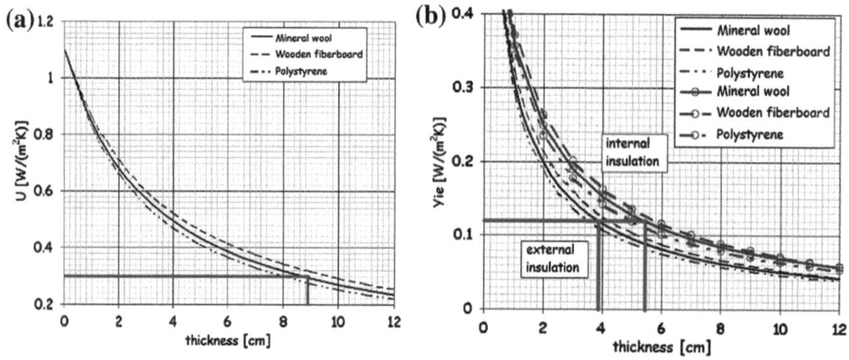

Fig. 8 **a** Thermal transmittance *U*. **b** Dynamic transmittance Y_{ie} according to the insulation thickness (internal and external insulation layer)

Table 16 Insulation thickness (external/internal)

Defined value	External insulation (cm)	Internal insulation (cm)
$U = 0.30$ W/(m² K)	7	7
$Y_{ie} = 0.12$ W/(m² K)	4	6

case, the amount of condensation is moderate and lower than the maximum values for the materials involved. In case of internal insulation, the climatic conditions of Rome do not cause condensation at all; in Madrid, condensation occurs, but it satisfies the two criteria for interstitial condensation (acceptable maximum

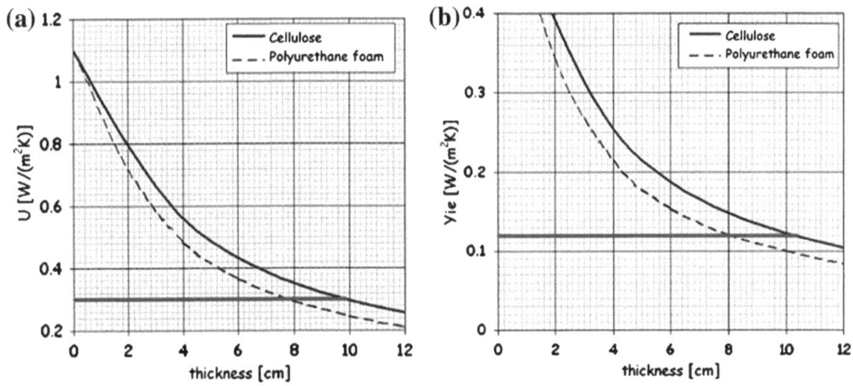

Fig. 9 **a** Thermal transmittance U. **b** Dynamic transmittance Y_{ie} according to the insulation thickness (air layer)

condensate quantity and complete evaporation). In the other climatic conditions (Milan, Paris, Brussels, and Hamburg), the wall is not verified.

The calculations show that the wall is free of condensation for all the climatic conditions considered, if the air cavity is filled with cellulose or polyurethane foam.

9 Case Study: Evaluation of the Envelope Incidence in Terms of Energy Performance

In order to estimate the importance of the envelope in building refurbishment, a set of simulations is carried out considering a case study. In particular, one flat of a typical construction built in the 1970s is adopted (Fig. 10).

The case study is a single dwelling in an intermediate level (adiabatic floor and ceiling). The main geometrical features are reported in Table 17.

The single glass windows have a wooden frame and a thermal transmittance of 3.44 W/(m^2 K). In order to investigate the different hygrothermal behaviours of some representative wall typologies described in Sect. 6, three opaque envelope hypotheses were defined according to the age of the analysed construction, as reported in Table 18.

The energy performance of the base case (without thermal insulation) was assessed according to the quasi-steady-state method presented in Sect. 6, referring to three different climatic conditions (Hamburg, Madrid, Milan).

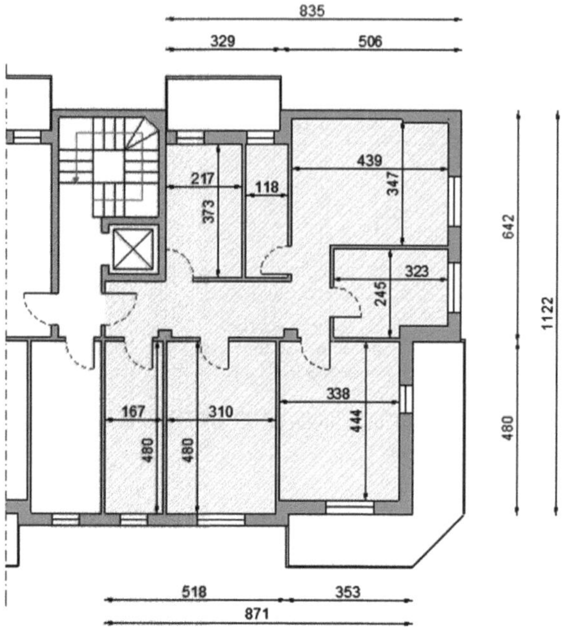

Fig. 10 Case study—general plan

Table 17 Building features

Internal floor area (m^2)	61.49
Overall conditioned volume (m^3)	260.12
Envelope surface/conditioned volume ratio (m^{-1})	0.46
Opaque envelope (m^2)	64.60
Transparent/opaque envelope ratio	0.21

Table 18 Case study—wall types and corresponding envelope winter energy needs

Wall type	U [W/(m^2 K)]	Y_{ie} [W/(m^2 K)]	$EP_{H,env}$ [kWh/(m^2 year)]		
			Hamburg	Madrid	Milan
10	1.16	0.656	185.58	97.74	157.47
12	1.098	0.594	152.53	76.49	127.88
17	1.626	0.599	200.16	107.14	170.52

9.1 Refurbishment Interventions

According to each building typology, a set of envelope renewal interventions is defined in order to evaluate the improvement of the energy performance. In particular, the four insulation materials reported in Table 19 are adopted, considering three different thicknesses (4–8–12 cm).

Table 19 Hygrothermal properties of the applied insulations

Material	[W/(m K)]	[kg/(m s Pa)]	ρ (kg/m^3)
Mineral wall	0.036	193×10^{-12}	40
Wooden fibreboard	0.040	97×10^{-12}	110
Polystyrene	0.033	1.3×10^{-12}	35
High-performance polyurethane	0.027	10.6×10^{-12}	35

Table 20 Percentage of improvement with the insulated walls

		Mineral wool			Wooden fibreboard			Polystyrene			Polyurethane		
L (cm)		4	8	12	4	8	12	4	8	12	4	8	12
10	Hamburg	35.9	43.7	47.1	34.5	42.7	46.3	37.0	44.5	47.8	41.2	47.5	50.0
	Madrid	43.5	52.8	56.8	41.9	51.6	55.9	44.8	53.7	57.5	49.5	56.8	59.8
	Milan	37.8	46.0	49.6	36.4	45.0	48.8	39.0	46.9	50.3	43.4	50.0	52.6
12	Hamburg	25.5	32.9	36.4	24.3	31.9	35.6	26.5	33.7	37.1	30.4	36.7	39.5
	Madrid	32.2	41.4	45.7	30.7	40.1	44.7	33.5	42.4	46.6	38.4	46.1	49.5
	Milan	27.1	35.0	38.8	25.9	33.9	37.9	28.2	35.9	39.5	32.4	39.1	42.0
17	Hamburg	39.7	47.5	50.8	38.3	46.5	50.0	40.8	48.2	51.4	45.0	51.1	53.5
	Madrid	47.5	56.6	60.4	45.4	55.0	59.1	48.8	57.5	61.1	53.7	60.8	63.5
	Milan	41.7	49.9	53.3	40.1	48.7	52.4	42.9	50.7	54.0	47.3	53.6	56.2

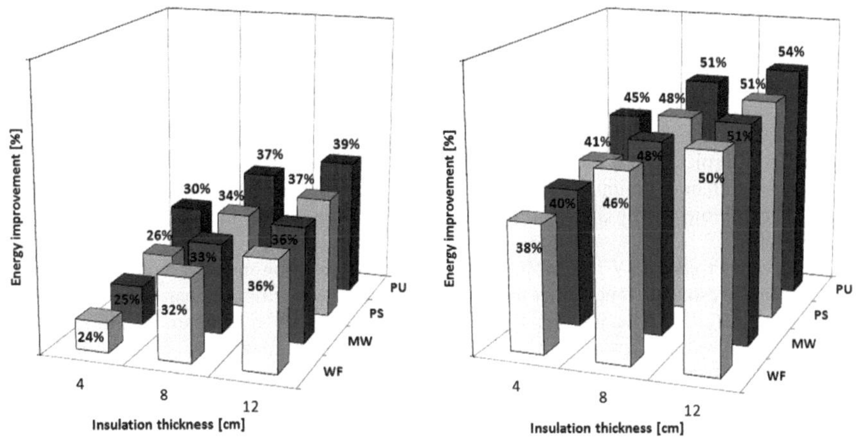

Fig. 11 Results for 12 and 17 walls (Hamburg). WF = wooden fibreboard. MW = mineral wool. PS = Polystyrene. PU = Polyurethane

In Table 20, the comparison of the improvement percentage in terms of envelope energy performance is presented, according to the kind of insulation and its thickness *L*, the climatic conditions, and the wall type that is investigated.

In general, it is possible to reach an average reduction, which accounts for 30–40 %, in comparison with the initial energy needs for the envelope, even with

4 cm of insulation. These values increase for the high-performance polyurethane, as it is expected, since it has a lower value of thermal conductivity.

It is apparent that for the wall 12, which presents the lowest value of thermal transmittance, the effect of the energy conservation measures is less significant than for 10 and 17 walls.

In particular, in Fig. 11, the percentage values for the wall types 12 and 17 are shown for the Hamburg climatic conditions. The differences related to the insulation types are negligible, except for the high-performance polyurethane, which is more effective than other materials. The main differences between the energy improvements are mainly related to the initial high thermal transmittance.

In fact, the energy refurbishment of low-performance buildings is more effective than in other cases. On the other hand, the less are the initial energy needs of the building, the more difficult is increasing the energy saving.

References

1. Corrado V, Ballarini I, Corgnati S, Talà N (2010) Use of building typologies for energy performance assessment of national building stocks, existent experiences in European countries and common approach, First TABULA Synthesis Report. Available from: http://www.building-typology.eu/existent-concepts.html
2. Magrini A, Magnani L, Pernetti R (2012) The effort to bring existing buildings towards the A class: a discussion on the application of calculation methodologies. Appl Energy 97:438–450
3. Magrini A, Magnani L, Pernetti R (2011) Analysis of the thermal control system incidence on the energy performance in building refurbishment strategies. 48° AiCARR international conference: energy refurbishment of existing buildings: which solutions for an integrated system, envelope, plant, control. Baveno, 22–23 September 2011
4. European Project EPISCOPE (2013–2016) Energy performance indicator tracking schemes for the continuous optimisation of refurbishment processes in European housing stocks
5. European Project Cost C16 (2003–2007) Improving the quality of existing urban building envelopes
6. European Project INVESTIMMO (2001–2004) A decision-making tool for long-term efficient investment strategies in housing maintenance and refurbishment
7. Magrini A (2013) Soluzioni per l'isolamento termico di edifici esistenti. Esempi di analisi termica e verifica igrometrica delle pareti (Thermal insulation of existing buildings. Examples of thermal analysis and hygrometric assessment of walls). EPC Editore, Rome

Reference Standards

8. CEN: Hygrothermal performance of building components and building elements: internal surface temperature to avoid critical surface humidity and interstitial condensation—calculation methods, EN ISO 13788:2013, European Committee for Standardization
9. CEN: Building materials and products: hygrothermal properties, Tabulated design values and procedures for determining declared and design thermal values, EN ISO 10456: 2008, European Committee for Standardization
10. CEN: Building components and building elements—Thermal resistance and thermal transmittance, EN ISO 6946: 2008, European Committee for Standardization

11. CEN: Energy performance of buildings—calculation of energy use for space heating and cooling, EN ISO 13790:2008, European Committee for Standardization
12. CEN: Thermal bridges in building construction—heat flows and surface temperatures—detailed calculations. EN ISO 10211: 2008, European Committee for Standardization
13. CEN: Thermal performance of building components—dynamic thermal characteristics—calculation methods, EN ISO 13786:2008, European Committee for Standardization
14. European Union: Directive 2002/91/EC of the European Parliament and of the Council of December 16th, 2002 on the energy performance of buildings. Official Journal of the European Communities, 4 January 2003
15. European Union: Directive 2010/31/EU of the European Parliament and of the Council of May 19th, 2010 on the energy performance of buildings (recast). Official Journal of the European Union, 18 June 2010

Transparent Building Envelope: Windows and Shading Devices Typologies for Energy Efficiency Refurbishments

G. Cellai, C. Carletti, F. Sciurpi and S. Secchi

Abstract Main typologies of windows, typical of the existing buildings, and innovative solutions, special glasses, and shading devices (fixed shading, mobile shading, roller blinds, and curtains) are described and assessed. The windows and solar shadings' appropriate choices are evaluated on the basis of a case study. For each of these solutions, thermal efficiency, natural lighting, and acoustic performances have been assessed with appropriate calculation codes. Dynamic computational methods with a graphical interface are used (EnergyPlus, through the Design Builder interface, for energy simulations, RELUX to simulate natural lighting, and DISIA for the acoustic simulations). Four representative climatic datasets corresponding to various locations (Berlin, Milan, Florence, and Athens) were considered. Appropriate performance indicators (defined by regulations or conventionally applied in science) have been identified in order to analyze performances and to evaluate different strategies for the achievements of energy efficiency and of comfortable environments: Q_{sw} (winter solar gains), θ_o (operative temperature), F_w (reduction factor of winter solar gains), DF (average daylight factor), UDI (useful daylight illuminance), daylight uniformity, $D_{2m,\ nTw}$ (acoustic insulation of facade normalized with respect to the reverberation time), and Δl_{fs} (sound pressure level difference due the façade shape). Starting from the performance evaluation of existing buildings, according to a logic implementation of consequential performance, this study provides for the assessment of different phases: the first interventions (phase A), replacement of existing windows with

G. Cellai (✉) · C. Carletti · F. Sciurpi · S. Secchi
Department of Industrial Engineering, University of Florence, Florence, Italy
e-mail: gianfranco.cellai@unifi.it

C. Carletti
e-mail: cristina.carletti@unifi.it

F. Sciurpi
e-mail: fabio.sciurpi@unifi.it

S. Secchi
e-mail: simone.secchi@unifi.it

A. Magrini (ed.), *Building Refurbishment for Energy Performance*, Green Energy and Technology, DOI: 10.1007/978-3-319-03074-6_2,

other high-energy performance ones (phase B), adaptation of the thermal transmittance of opaque envelope to national limits (phase C), and introduction to solar systems and solar control glasses (phase D). Then, the effect of screens and windows on the reduction in the thermal loads in the summer season and on the thermal comfort has been assessed, together with the influence on visual and acoustic comfort of different configurations of windows and shielding. Finally, a comprehensive evaluation on the aspects of energy consumption, natural lighting, acoustic comfort, and technical feasibility (TF) is carried out.

Symbols

A	Area (m^2)
A_g, A_f	Glass and frame area (m^2)
$D_{2m,nTw}$	Standardized façade sound level difference (dB)
DF	Daylight factor (%)
E	East orientation
E_{int}, E_{ext}	Indoor and outdoor illuminances (lux)
$E_{\mathrm{min}}/E_{\mathrm{m}}$	Daylight uniformity (–)
g	Solar factor (%)
I_{sol}	Solar radiation (W/m^2)
N	North orientation
Qs_w	Winter solar gains (kWh)
Qs_s	Summer solar gains (kWh)
θ_{o}	Operative temperature (°C)
F_w	Reduction factor of winter solar gains (%)
F_s	Reduction factor of summer solar gains (%)
Ra	Color-rendering index (–)
R_w	Rating of sound reduction index (dB)
S	South orientation
SC	Shading coefficient (%)
U	Thermal transmittance [W/(m^2 K)]
UDI	Useful daylight illuminance (–)
U_g	Thermal transmittance of the glass [W/(m^2 K)]
U_f	Thermal transmittance of the frame [W/(m^2 K)]
U_w	Thermal transmittance of the window [W/(m^2 K)]
l_g	Total perimeter of the glazing (m)
Ψ	Linear thermal transmittance [W/(m K)]
W	West orientation
Y_{IE}	Periodic thermal transmittance [W/(m^2 K)]
ΔL_{fs}	Façade shape level difference (dB)
φ	Phase shift
Ψ_g	Linear thermal transmittance of glass [W/(m K)]
α	Noise absorption coefficient (–)
λ	Thermal conductivity [W/(m K)]
θ_{db}	Dry bulb temperature (°C)

τ_v Light transmittance (%)
τ_λ Spectral light transmittance (%)

1 Generalities

The relation between window openings and outdoor calls for three specific needs:

1. heat flow control through components with a low thermal inertia capacity;
2. protection from solar radiation;
3. visual connection between the envelope and the outside and therefore a satisfactory level of natural light.

The first requirement, strictly related to energy efficiency, has pushed forward the development of material components with low transmittance values (thermal transmittance value U_W): their improved performances are producing positive outcomes in terms of thermal and acoustic comfort.

The second requirement deals with the light control through the glass panes (with total solar transmittance g value and light transmittance τ_v) and with the application of screen devices to protect against solar radiation and to ensure occupants' comfort.

The third need is prominently centered on the visual comfort (VC) demand.

The replacement of windows, aimed to energy refurbishment, represents the kind of intervention that brings the most efficient cost/performance ratio, often promoted by tax incentives; this kind of solution is also practicable by simply substituting the existing glazing with new ones, maintaining the same frame.

2 The Thermal Performance

Three components define the thermal transmittance of a window (U_W): the glass panes, the frame (fixed or operable), and the spacer between panes (multiglazed windows).

The following heat exchange process combinations must be taken into account (Fig. 1):

- Convective and radiative heat transfer between the outer and the inner adjacent surfaces;
- Conductive, convective, and radiative heat transfer within the cavities of the window itself (double-glazed or double-framed windows).

The thermal transmittance value can be calculated as follows [27]:

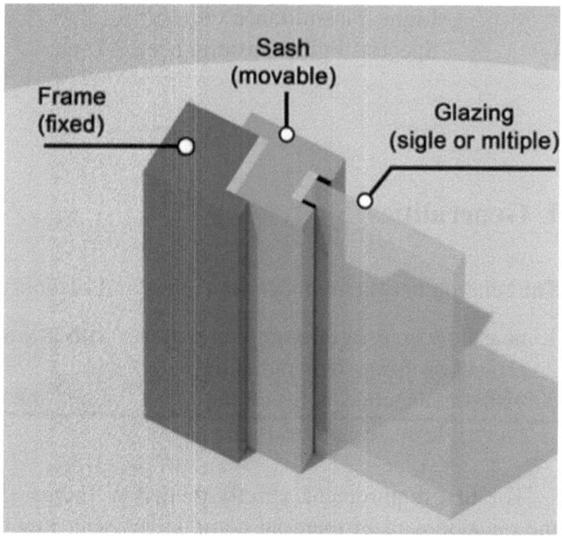

Fig. 1 Single window scheme

$$U_W = \left(A_g U_g + A_f U_f + \Psi_g l_g\right) / \left(A_g + A_f\right) \left[\mathrm{W}/(\mathrm{m}^2\,\mathrm{K})\right] \quad (1)$$

where

U_g Thermal transmittance of the glass [W/(m^2 K)]
U_f Thermal transmittance of the frame [W/(m^2 K)]
A_g, A_f Glass and frame area (m^2)
l_g Perimeter of visible glass edge (m)
Ψ_g Linear thermal transmittance due to the combined thermal effects of glazing, spacer, and frame [W/(m K)]

For a conservative approach, the computation of U_W is proceeded under the common practice of neglecting the shield film layer, if present, handling it as inactive (this does not apply, however, to the procedure of irradiance considering solar gain calculation).

For double-glazing panes bonded around the perimeter of the spacers, the cavity holding air, performing a low value of thermal conductivity ($\lambda \cong 0.025$ W/(m K)), plays a decisive contribution in terms of thermal resistance improvement.

Further on, the reduction in conductive and convective heat transfer is possible, thanks to the development of several solutions, such as in Fig. 2:

- the application of gas layer with a thermal conductivity lower than the air conductivity (for instance, argon and krypton gases);
- the introduction of coating film over the glass panes with a consequent emissivity reduction (low-emission glazing);
- the addition of interspaces splitting with multiglazing systems;
- the adoption of spacers with low thermal conductivity material components.

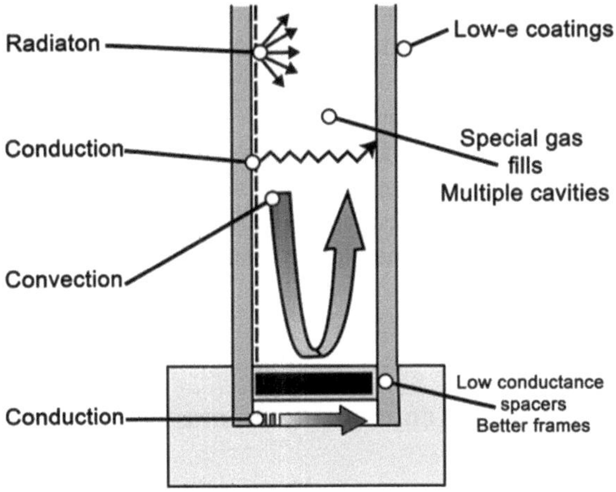

Fig. 2 Thermal transmittance conditions and improvement actions

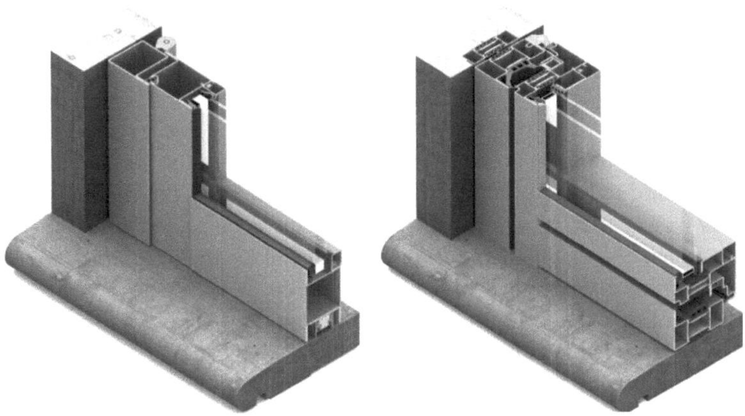

Fig. 3 Examples of typical aluminum frames without (*left*) and with (*right*) thermal break (*strips of polyamide*) (*Company METRA*)

The ideal limit occurs with the adoption of vacuum cavities between glass panes: the heat exchange is reduced to just radiation and the cavity reaches the maximum value of thermal resistance equal to approximately 0.276 m^2 K/W, with a temperature difference between glass panes of about 10 °C [7].

The thermal transmittance of the glazing U_g depends on the number, the thickness, and the interspace of glazing. U_g values are also depending on emissivity and nature of the gas in the cavity, and they can reach about 0.3–0.5 W/(m^2 K), congruently with the "zero-energy house" achievements.

The frame, along with the glass panes, constitutes the other essential components of the window; it can be made out of wood, aluminum, or metal with thermal break (Fig. 3), PVC, or mixed materials.

Historically, for doors and windows, the most common material is wood, with frame thicknesses between 50 and 60 mm, and average transmittance U_f of about 1.9–2.3 W/(m^2 K) (generally wood frames perform higher transmittance values than metal frames with thermal break).

Nowadays, to improve the performance and the durability of wooden frames, mixed solutions have been developed with aluminum on the outer side and wood on the inner side, with thermal insulation components and U_f values between 1.0 and 0.6 W/(m^2 K) (about six times lower than the values of metal frames with thermal break) [9].

3 The Radiative and Acoustic Performance

Common glass panes are transparent to solar radiation in the range of wavelengths from ultraviolet to the near infrared (from 0.3 to 2.5 μm), with a maximum peak in the visible range (about 42 % of the solar energy is emitted in the range of 0.38–0.74 μm wavelengths).

In order to control the transmission of solar radiation, a protective coating can be applied over the glass surfaces, improving the light transmittance τ_v (%) and the solar factor g (%).

Both parameters represent the average values of the energy ratio transmitted through the glass and the normally incident energy over the surface in the spectrum of standard radiation [26]. In Anglo-Saxon countries, the shading coefficient SC (%) is widely used as a valid alternative to the solar factor g. It is the ratio of the radiant energy that penetrates through the glass to the energy transmitted through a common clear glass of 3 mm thickness.

The control of radiation, primarily, is performed by increasing the capacity of reflection in the visible range, but this action is likely going to alter color perception, expressed by the general color-rendering index Ra.

The index Ra can reach the maximum value of 100 for glasses whose spectral transmission factor τ_λ is constant in the visible spectral range. For common glazing, $Ra \cong 98$ and $g \cong 0.89$.

In the field of environmental control practices pertaining to lighting design, $Ra > 90$ is featuring a very good yield, while values of $Ra \geq 80$ indicate a yield of acceptable color.

The chromaticity of glass panes, in relation to comfort, shows a comfort acceptability falling below 85 % of occupants when τ_v is reaching 38 % with cloud cover and 25 % with clear sky.

To ensure an effective protection against solar radiation, a glass must have a g value between 15 and 20 %; however, this implies a high reduction in τ_v, with consequent worsening of natural lighting, and sometimes, it can result in a Ra shortcoming [2].

The solar radiation effects could be reduced using special coated glasses, such as:

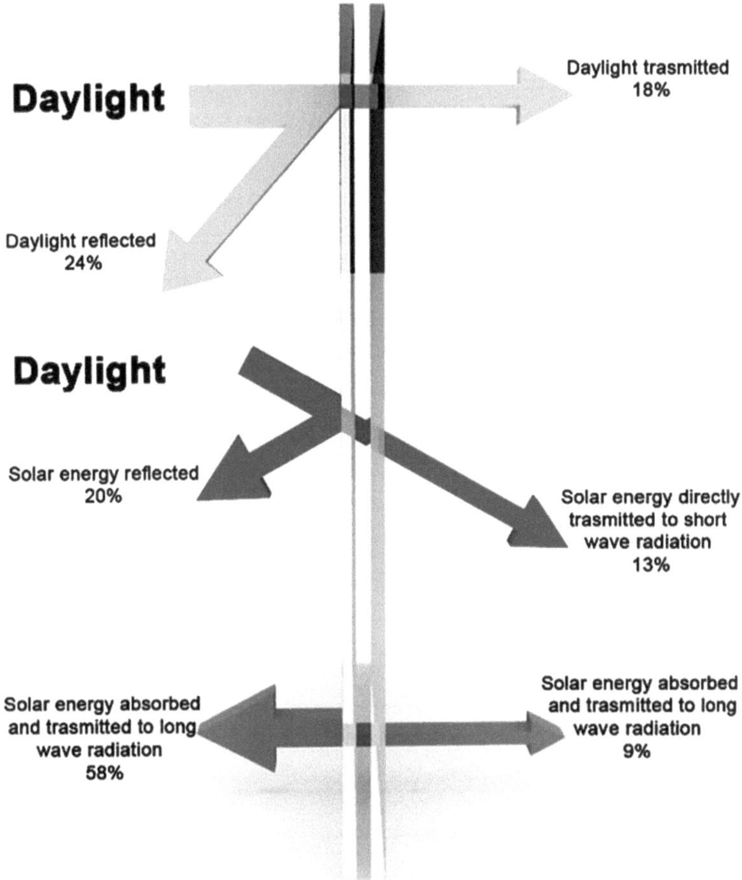

Fig. 4 Example of solar reflective glass window 6–12–6 mm, with $U_g = 2.7$ W/(m^2 K), light transmittance $\tau_v = 18$ %, solar factor $g = 22$ %, and $SC = 25$ % [17]

1. sunscreen-reflective glass (Figs. 4 and 5);
2. glass for thermal insulation—low emission;
3. low—emission glass and reflective glass(Fig. 6).

Along with coatings, the proper selection of tinted glass plays a key role in architectural design. The internal chromatic perception is the response to the light transmitted through the ranging variety of stained panes: the most common are gray, bronze, blue, and green (Fig. 5); those colors do not excessively alter occupants' perception ($Ra \geq 90$).

Moreover, a glass pane that creates a brighter light effect in rooms is more suitable (such the kind in bronze color does); usually, in hot climatic regions, cold colors are preferred and vice versa in the cold ones [3].

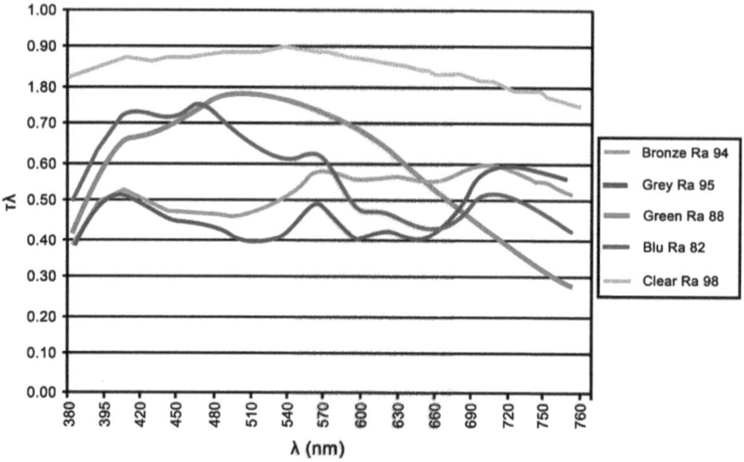

Fig. 5 Tinted glasses—spectral trend τ_λ and color-rendering index *Ra*

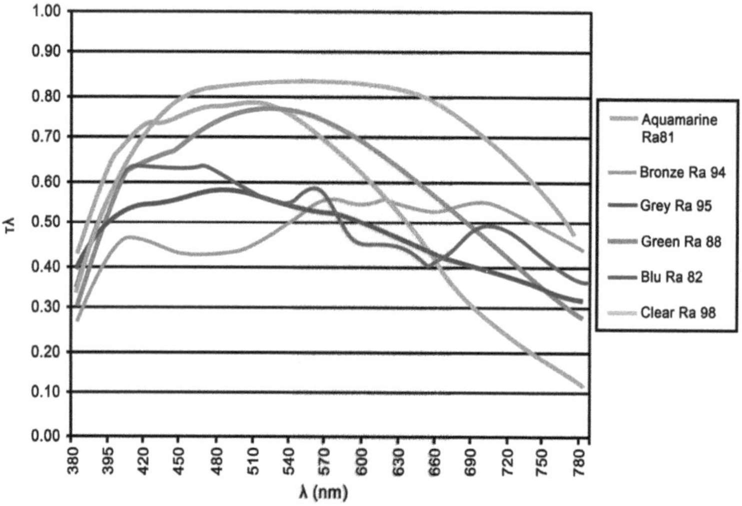

Fig. 6 Selective coated glasses—spectral trend τ_λ and color-rendering index *Ra*

In the making of choices for glass types, therefore, being aware of transitional effects from winter to summer, façade exposures, shading effects from the surrounding, etc., would be useful. Emerging technologies aim to develop glass panes with dynamic optical properties, such as the smart windows, consisting of electrochromic materials (ECWS), liquid crystal devices, or suspended particles [10]. They are capable of varying the radiative performance in function of the outside weather and the inside conditions, cutting down energy costs by 10–45 % on lighting and 5–15 % for air-conditioning in summer, up to a maximum of 20 %, compared to more conventional control systems for solar radiation.

The electrochromic glass may have a τ_v factor from 0.03 to 0.75, with thermal transmittance of $1.6 < U_g < 0.5$ W/(m^2 K) [10].

Natural light changes during the day and with the changing seasons; thus, the levels of internal lighting and external lighting are instantaneous values that are constantly changing, especially in the presence of variable sky. Consequently, it is not possible to prescribe absolute limit values of natural lighting. In addition, in clear sky conditions, the assessments should be conducted under detailed computational method procedures.

For simplicity, many regulatory and legislative codes refer to the relation between the internal lighting and external lighting at the same time; the performance indicator, the "daylight factor" (DF), is used to evaluate and express the punctual nodes of luminance levels of zones under an overcast sky, and it is defined as the ratio of internal (E_{int}) illuminance to external (E_{ext}) illuminance:

$$\text{DF} = \frac{E_{\text{int}}}{E_{\text{ext}}}$$

Since the percentage of DF varies in every point of a given environment, usually it refers to its average value. The greater the homogeneous distribution of light, the greater the internal level of comfort.

According to EN ISO 12464-1 [28], it is desirable that the ratio of minimum luminance to average luminance is ≥ 0.7 in the area of the visual task and ≥ 0.5 in the immediate surrounding area to that of the task.

The light distribution (gradient lighting) also presents spatial variations: the level of natural light decreases rapidly with increasing distance from the window (Fig. 7); the interior surfaces, including the furniture, can produce strong contrasts or reduce differences in brightness. A side sourced light presents a very different gradient of illumination from a zenithal light source, characterized by a relatively uniform distribution. An external obstacle can hinder the internal lighting.

One of the parameters used for the analysis of illuminance levels for inside spaces is the useful daylight illuminance (UDI) [13, 14]; the UDI has been defined in order to support the analysis of illumination levels by natural light, based on hourly meteorological climate data for the period of 1 year, to determine how many hours the level of natural lighting is within the range 100–2,000 lx deemed satisfactory by the users for an adequate VC in natural light conditions.

Below 100 lx, illumination values of natural light are considered insufficient to satisfy basic visual tasks. They represent a negligible contribution toward energy efficiency. Natural light levels in the range 100–500 lx are effective for many visual tasks and offer a good contribution in terms of energy savings. The illuminance values of natural light in the range 500–2,000 lx are considered satisfactory for all the visual tasks, while illuminance values >2,000 lx, in most applications, cause visual discomfort and rising of temperature.

In the evaluation of solar shading solutions, the UDI is used to evaluate the lighting properties of shielding, with "clear sky" conditions corresponding to the

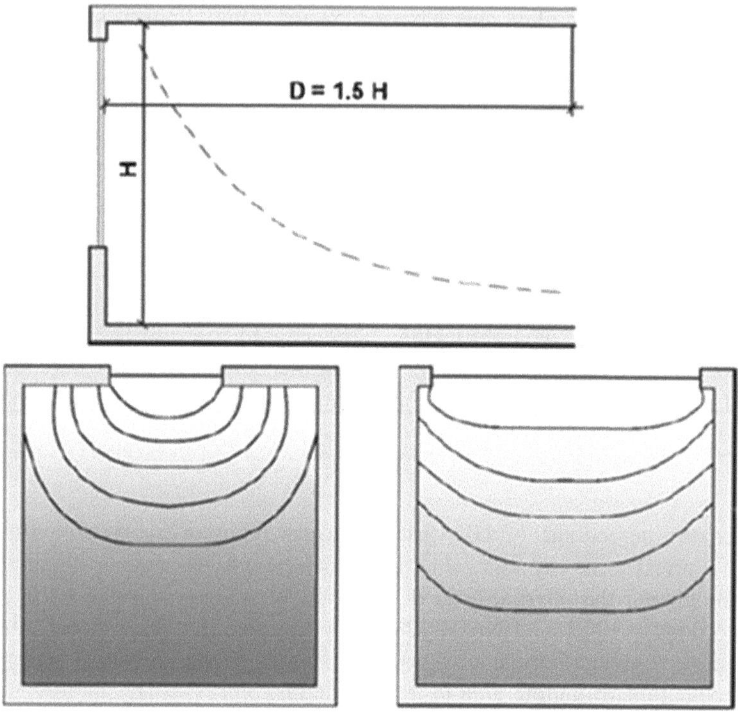

Fig. 7 Gradient lighting depends on the height (H) and the width of the window

model developed by CIE, which provides a luminance distribution as a function of the position of the sun at given latitude [18, 19].

Another fundamental aspect is given by acoustic comfort specifications, which represent a design parameter setting in all European countries. Acoustic performances are of great importance in the design of residential, considering that, very often, they are the critical part of the partitions and closures and they have to meet standards of comfort expectancies and building codes and regulations. The noise, in fact, may have effect on people's health and consequently economic implications.

With regard to acoustic performance of windows and shading devices, they are expressed by the index of evaluation of the sound reduction R_w (dB) [5]. Another fundamental aspect to ensure the expected performance is related to the tightness of the frame [21], which must be as highest as possible; otherwise, the penalty may also be of several dB.

The acoustic performance of a window is conditioned in order of importance by [4, 24]:

- the type of glazing (single glazing or laminated);
- the number and thickness of glasses (mass of the component);
- the tightness of the frame (type of seals and number of beats).

The installation procedures need proper attention, as they have to be performed, for example, without creating sound bridges at the frame–masonry junctions.

Summarizing as a reference of the above-covered contents, the following performances are listed below:

- solar factor $g \leq 0.50$;
- light transmittance $0.38 < \tau_v < 0.65$;
- color-rendering index $Ra \geq 90$;
- thermal transmittance $U_g \leq 1.3$ W/(m^2 K);
- rating of sound reduction index R_w of the window ≥40 dB.

4 Windows' Typologies

The general term of "external openings" widely refers to several technical elements such as fixed windows and operable windows (by single-hung or double-hung sash, horizontal sliding sash, awning, hopper, tilt and slide, and tilt and turn) jalousie window, clerestory, roof lantern, skylights, French doors etc.

According to the definition given by the standard [25], the window is a building component for closing an opening in a wall or pitched roof that will admit light and may provide ventilation.

The current production of windows can be classified according to:

- type of opening;
- type of frame materials;
- type of glass;
- type of spacers between the panes.

Each of the above typology components can lead to issues, mostly in terms of acoustic and thermal bridges; both of these aspects are inherent to the technology implied in the manufacturing and in the installation procedures.

4.1 Installation

The installation procedures of windows must attend to the main requirements of:

- control of thermal bridges through the window framing and the wall system;
- control of sound transmission through the window framing and the wall system.

Both of these requirements have to be fulfilled; otherwise, good energy-saving windows may have their performance diminished as a result of a poorly proceeded installation.

4.2 Thermal Bridges

An important reference supporting the first requirement is the EN ISO 14683 [29].

Looking at the various typologies of thermal bridges, the most critical conditions can be identified as those having a maximum value of linear thermal transmittance Ψ. These are constituted by:

- window positioned on the outer side with Ψ between 0.8 and 1 W/(m K);
- window positioned at the center of the wall with Ψ between 0.6 and 1 W/(m K);
- window frame positioned on the inner side of the masonry with Ψ between 0.8 and 1 W/(m K).

The best solutions with values $\Psi \leq 0.20$ W/(m K) are those with the frame resting directly on the insulating layer.

4.3 Sound Control

The situation in terms of sound control is more detailed and complex, considering the crucial importance to achieve sound capabilities of the insulation layer concerted with the choice of obscuration (blinds or shutters). Also, the inevitable uncertainties and worsening factors need to be considered, related to the construction conditions of the various components of the façade. In this regard, building guidelines that provide specific indications for proper installation (for example, the Italian standard [30]) are useful references.

Once the type of windows has been identified, the next task is to define some details on the wall opening that are critical to keep the building envelope sealed from water, air, and heat transfer and also to maintain high sound insulation performances.

In particular, the wall opening, enforced by the introduction of a wall curb sill extruded over the jamb perimeter (Fig. 8 left), may reach an acoustic performance better than the simple frame system lying flashed along the exterior wall casing (Fig. 8 right), through which sound waves can propagate more easily toward the internal environment.

A careful handling of dimensional tolerances is an unavoidable precaution to prevent serious consequences during the window installation process. It is recommended a tolerance with at least 5 mm per side between window frame and wall opening, variable in function of frame materials, solar radiation absorption capabilities, frame color and size (PVC in dark color is quite sensitive, for instance, to thermal expansion).

Bad junctions between masonry and window, small holes or poor realizations of the attack, can affect the overall result, with reduction in more than 10 dB of sound insulation.

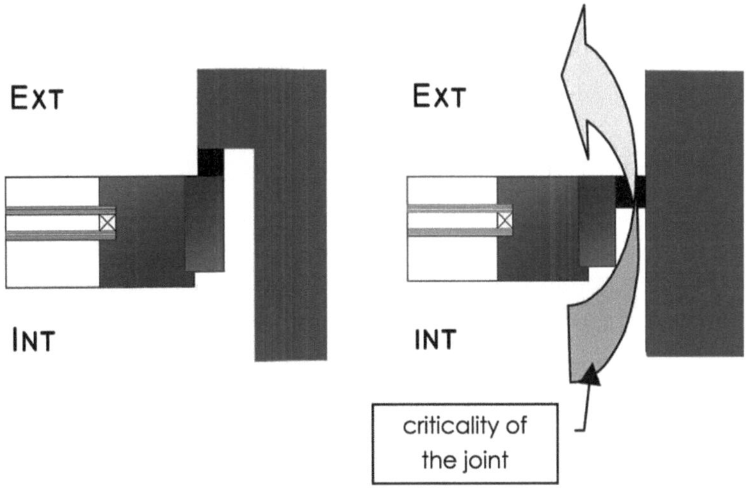

Fig. 8 Example of a wall curb sill extruded over the jamb of the window frame (*left*) and window frame flashed along the exterior wall casing

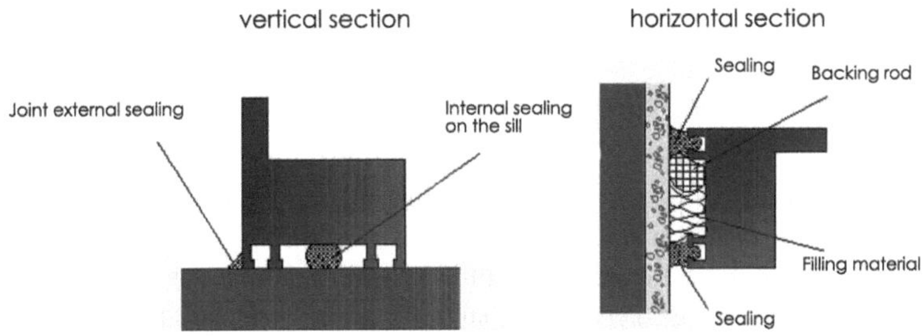

Fig. 9 Example of realization of a coupling in light

In case of window shutters lying on the middle line of the wall section (Figs. 9, 10), a bead of sealant should be applied on the three shoulders of the window opening and the sill, making sure to connect them. Once fixed the window frame into the wall compartments, it is necessary to perform the operation of filling the joint with expanding material and to seal properly the inner part of the joint with sealant; then, the interstice between the masonry and the window frame must be sealed by proper material.

Another important element is the interface frame since that, if poorly executed, can compromise the overall operation of the window.

Poorly executed installation of the glass pane interfacings can also produce issues related to sensible deformations over the frame profiles (for instance, the excessive weight of the glass plates can reduce the air tightness of the frame and therefore affects the sound insulation performance).

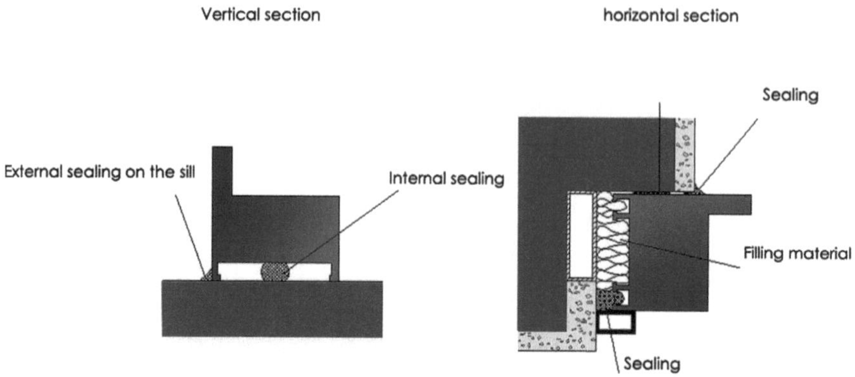

Fig. 10 Example of construction of a joint in abutment

In this respect, the proper sizing and positioning of the blocks are important. These procedures contribute to relieve the weight of the glass plates over the frame and to keep them in the right position, avoiding movements of the sheets of glass chamber.

4.4 Window Rolling Shutter Casing

Particular attention has to be paid to the installation of window rolling shutter casing, since it may cause several weaknesses in terms of sound insulation.

The installation must ensure that both the casing box and the maintenance door are secured by durable gaskets with a suitable grade of elasticity.

In the case of prefabricated rolling shutter casing, the material space between the wall and the prefabricated block on the shoulders, laterally and above the casing itself, must be filled carefully with mortar or expanding materials.

Depending on the type of opening, windows may have major or minor shortcomings in terms of thermal and acoustic performances (it should be also considered the opportunity given by hopper window and tilt and turn window).

Among the various types of windows, the ones that ensure the best possibility of sealing, and consequently the best thermal and acoustic performances, are those with one leaf, namely those that, for the same frame and surface, have the lowest perimeter of the stop.

The two doors' frames can have opening inward or outward and an opening in rotation around the two vertical side axes, therefore with a very critical stop perimeter than the one door window.

An alternative can be represented by the revolving door window, with an opening in rotation on the vertical axis or on the horizontal axis (horizontal hovering). When fitted with a locking mechanism, the ventilation is permitted without full opening.

Bottom hinged windows are essentially conceived for services or workplaces to allow ventilation; in today's residential buildings, they are often replaced by sash windows with a turn and tilt opening system.

For windows with important surface area opening, sliding horizontal doors are often adopted. These types of frames, which are difficult to seal, have been recently improved by the introduction of the sliding coplanar door windows that align on a track the two sides of the window. The coplanar sliding doors have a double system of opening: sliding door and hopper.

The so-called foldable windows, composed of multiple door components that can be folded, are much more critical, especially under the acoustic profile.

4.5 Frames

The frames can be classified according to the system of beaten into three categories:

- Windows with single beaten;
- Windows with double or triple stop beatens;
- Window frame with open joints.

The frames with single stop profile have a simple single seal, which has to guarantee air and water tightness; therefore, it is generally not reliable, especially in the presence of high external pressure, when the wing tends to inflect the frame itself causing the detachment of the gasket. At this purpose, the sealing of windows with double or triple stops should be preferred.

A further evolution consists of the introduction of an open joint between fixed frame and opening section, providing a capillary break, which prevents water seeping in and lodging in the joint. The central gasket allows to drain the water eventually penetrated inside, through the drain holes, using an equilibrium phenomenon of internal pressure to the external one.

The mechanical performances required for the casing are:

1. air tightness;
2. water tightness;
3. resistance to wind;
4. mechanical resistance.

The choice of the energy classes for exterior windows has to be performed considering the characteristics of the building and the climatic conditions of the specific environment. In addition, the performances must be appropriate to the size, type of windows, and levels of thermal and acoustic insulation required within the living spaces.

4.6 Permeability, Water Tightness, and Wind Resistance

Windows can be cataloged into four classes with regard to air permeability [20], into about 18 classes for their water tightness [22] and in seven classes with regard to wind resistance [23].

5 Solar Shading Typologies

The evolution of the environmental control techniques combined with the support of detailed computational software inquire to the designer a careful choice of shading devices in the building design, in order to ensure the consumption control during summer times and to provide comfort for the occupants [15].

In general, a screening system can be applied to the right-angle frame of the entire building or a portion of it, adding also a value to the renewal of the façade design, so that their application allows a new perception into existing buildings.

A well-conceived shading device must be able to maximize heat gains in winter conditions and to control the radiant heat in summer conditions, as well as to improve visual and acoustic comfort of the interior.

Accordingly, the effectiveness of sun protection of glass surfaces depends on different factors:

- characteristics of the screen materials and finishing (reflectance);
- solar shading solution (fixed or mobile). A fixed solar shading (canopies, balconies, frames, etc.) does not allow variation in energy responses; on the contrary, mobile shading devices permit, manually or by automated systems, to adapt to the sun path daily and yearly, due to a punctual control of the shading elements to ensure natural light maximum efficiency;
- screen positioning with respect to the frame (external, internal, intermediate). The outer shields are most effective, intercepting the incident solar radiation before the glass panes and preventing therefore the greenhouse effect. Furthermore, the placement of the shielding outside also allows to interact with the outer sound waves (for instance traffic). Thus, if properly designed, external shielding can help to significantly reduce the sound pressure incident on the façade;
- screen disposition, according to the façade exposition (parallel, orthogonal, horizontal, vertical, etc.), geographical location, and thermal loads. Often, the shield with vertical fixed elements is conceived for the areas facing east and west on which solar radiation affects the morning and late afternoon, with a lower height on the horizon profile. The system that places the fixed elements perpendicular to each other, called "grating", is one of the most suitable for shielding glass surfaces located at the east and west side in hot climates, but these elements are hardly applied in residential buildings and are most used in public and industrial ones, due to the strong architectural language they impose.

All the sun protection systems installed in front, or on the outside of the frame, without making the same body, are external sunscreens, by definition.

Solar shading device can be defined as a "screen attached to the outside of the wall that consists of several horizontal or vertical elements with the function of sun radiant energy mitigation."

Outdoor sunscreen solutions bring higher added value both in terms of architecture and in terms of economic performance; external shieldings, much more effective than internal ones, are usually more expensive and subject to maintenance, since they are permanently exposed to atmospheric agents. The type of material of screen components plays a fundamental role.

The blinds elements are mostly made out of extruded aluminum or galvanized steel and painted. There are also brise-soleil made out of other materials such as wood, brick, PVC, and copper. Whatever the material employed is, the device must ensure adequate operability and aesthetic value over time; in this regard, metals, properly treated and painted, have effectively replaced wood, more prone to deterioration.

The selection of external shielding components should take into account the outdoor weather conditions and the device wind resistance, since, under high wind loads, the system can suffer major damages. With regard to blade sections, there are several possible configurations: ellipsoidal, curved, gull wing, triangular, diamond-shaped, and rectangular (for wooden ones).

The several solutions for external shielding can be classified into four main product families:

- fixed shading;
- mobile shading;
- roller blinds;
- curtains.

5.1 Fixed Shading

These systems are commonly constituted by a shield of linear panels or slats, mounted in parallel on a fixed or adjustable frame, to form a pattern to intercept the solar radiation. Figure 11 shows some examples of fixed shading.

Sunscreens are installed in front of the window with preoriented blinds or blades, creating a kind of outer curtain. Generally, the blades are always geared according to the sun incidence of the hottest period of the year.

These systems may become a very important formal element in the project, if they are interpreted as a real envelope.

Fig. 11 Examples of fixed shading (from the *left* horizontal sunscreen, fixed overhang, grating and sunscreen, fixed blades)

5.2 Mobile Shading

This solution, featured to shield the building from solar radiation, by modifying the blinds or blades angle, allows to optimize the amount of natural light. Figure 12 shows some examples of mobile shading.

The devices that rely on blinds are installed horizontally to the façade, while those with blades can have a vertical application too. Vertical blades are smaller in size and can rotate by about 180°; it is a product used primarily for industrial application. The rotation of the blades allows to shield the radiation and to reflect it into the enclosure by adjusting the flow. This system, usually of significant dimensions, is lacking of the capability to eliminate all the shield obstruction, when not required, since the blades attached to their pivot are not packable.

Shadings with horizontal blades have larger dimensions to accomplish aesthetic purposes and to resist against the wind. These devices come with various section profiles (the most common is ellipsoidal), with large intersection, and can be automated by light sensors which allow a continuous variation according to the daily sun path.

The so-called Venetian blinds are very similar to the Venetian curtains: the main difference is in the size of the blinds. They consist of a cloth made out of painted aluminum planks or slats of various shapes and sizes hanging from a ladder tapes. The slats are driven through a mechanisms housed in the upper casing; along the sides, lateral guides or aluminum wires are provided. The key feature is the ability to top packaging into a very small space that favors their use even in residential building renovations, where there is not much available space and the option for façade inclusions does not meet the aesthetic expectations or the regulatory requirements.

Persian blinds consisting in sliding or folding doors, also with operable slats, are widely used in residential constructions and can be produced in different materials such as wood, aluminum, and PVC. The choice of suitable materials, in addition to aesthetic reasons, is also linked to maintenance needs.

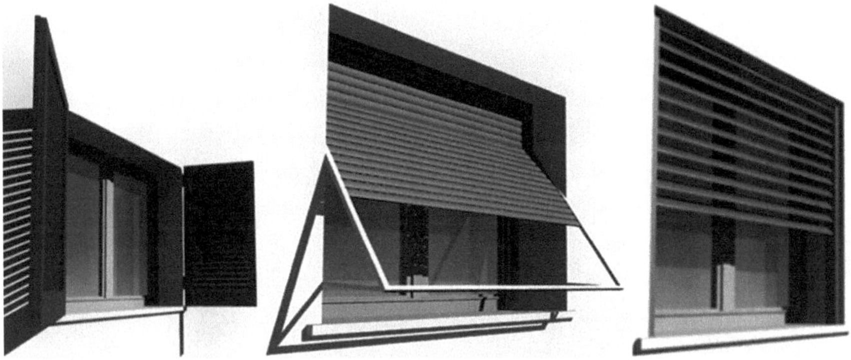

Fig. 12 Examples of mobile shading (from the *left* persian shutters, roller blinds, and Venetian blinds)

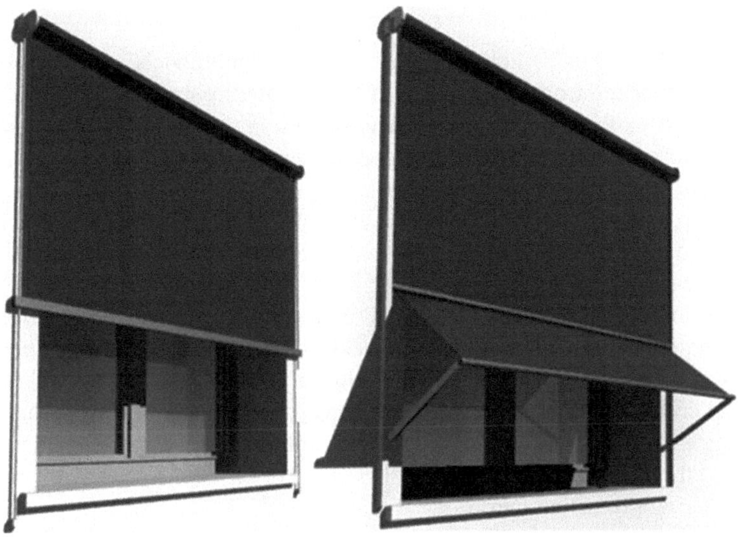

Fig. 13 Examples of roller blinds (*left* roller curtain; *right* sliding arm awning)

5.3 Roller Blinds

These systems are widely used, due to the simplicity of the mechanism (spring-roller-operated, gearbox, or engine that wraps around the curtains) and also due to volume-saving characteristics, (Fig. 13).

From a solar control point of view, the degree of response depends exclusively on type, color, and weight of the fabric used.

Fig. 14 Examples of curtains (*left* drop-arm awning, *right* tent canopy)

5.4 Curtains

This family of screens includes various typologies and models, which ensure the protection of façade from solar radiation with limited costs and flexibility. Being very exposed to the elements, raw materials of high quality and sophisticated finishing are used to maintain unchanged the aesthetic and functional characteristics of the system itself (Fig. 14).

The drop awnings are often visible on balconies or directly build into the façade system, installed vertically, with variable size, overhang in order to close the openings and shading the area that lying below them.

Curtains may also have a foldable and retractable structure and can be installed in horizontal or tilted plane up to 90°; in addition, this type of installation facilitates proper ventilation of the spaces below. Nowadays, the performance of the tent fabrics is crucial on market competition. These components must have the ability to mitigate solar radiation, durability, and waterproofing capability. The most common fabrics are made out of glass fiber or polyester yarn both coated with PVC for weathering protection purposes.

A variation is represented by fabrics screened by an undercoating of PVC film, in order to permanently close the wefts of the tissue. This type of textile provides a heavy and stiff coat and therefore is used only in highly demanding applications.

5.5 Internal Solar Shadings

The inner shields are less effective from an energy point of view than the outer ones; thus, they are usually added to them to further control solar radiation, daylighting, and glare and to ensure privacy for occupants. Moreover, their

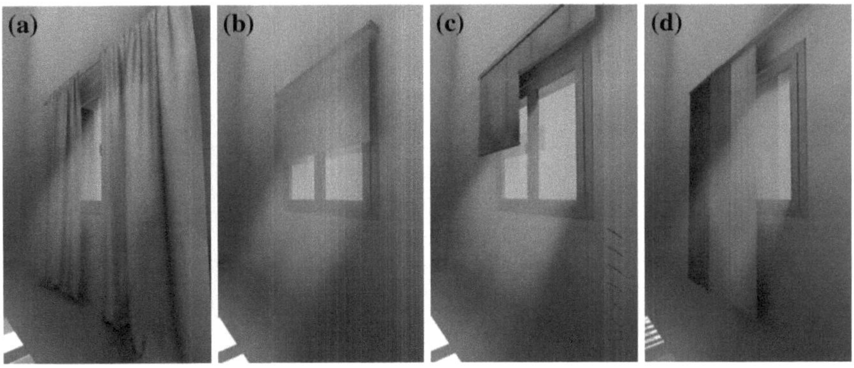

Fig. 15 Examples of internal blinds (from the *left* classic drapes, roller drapes, Venetian blinds, and vertical curtain)

capability to mitigate the sound pressure coming from the outside is negligible (Fig. 15). The principal products for internal application include:

1. translation systems (panel curtains and skylights);
2. chutes (roller blinds and pleated blinds);
3. systems to strip (horizontal Venetian blinds and vertical louvers).

5.6 Intermediate Solar Shading

One possible solution of combining the glazing with shielding systems derives from the existing Venetian blind systems, with smaller scale blinds to be placed in the cavity of the double glazing. This hybrid system provides a satisfactory level of solar radiation control and represents an efficient alternative to those previously described.

The system ensures an adjustable filter to the entrance of sunlight: the amount of light can be adjusted from 80 % total obscuration and instantly adjusting the brightness depending on the demand (Fig. 16).

The blinds are mounted within two panes of glass, and their scrolling takes place in a sealed chamber. This feature ensures total protection against dust and weather. The solution presents durability and maintenance issues.

There are many versions of this system, each with its own mechanical solution, aimed to solve the handling without compromising the seal of the glazing panel (mostly there is a magnetic slide system and motorized system).

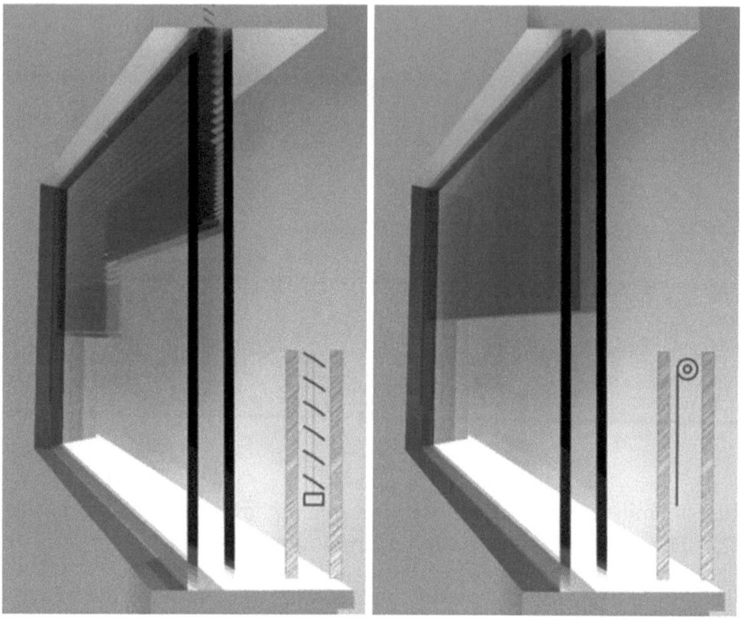

Fig. 16 Examples of integrated screens (from the *left* Venetian blind and roller blind)

6 Integrated Solutions Applied to a Case Study on Existing Buildings

The evaluation of different strategies for upgrading the energy efficiency and performance over various types of shielding needs to be explained by a case study. A typical room was taken into account, with features and dimensions representative of typical post-World War II Italian residential architecture (Figs. 17, 18).

Different strategies concerning windows and solar shadings have been applied; for each one, thermal, daylighting, and acoustic performances were assessed with appropriate calculation codes.

6.1 Case Study Description: Significant Parameters

For the purposes of the analysis, detailed computational methods working in dynamic regime and featuring a graphical interface are considered. These software applications are the following: EnergyPlus (through the Design Builder interface) for energy simulations [12], RELUX [16] to simulate natural lighting, and DISIA for the acoustic simulations [11].

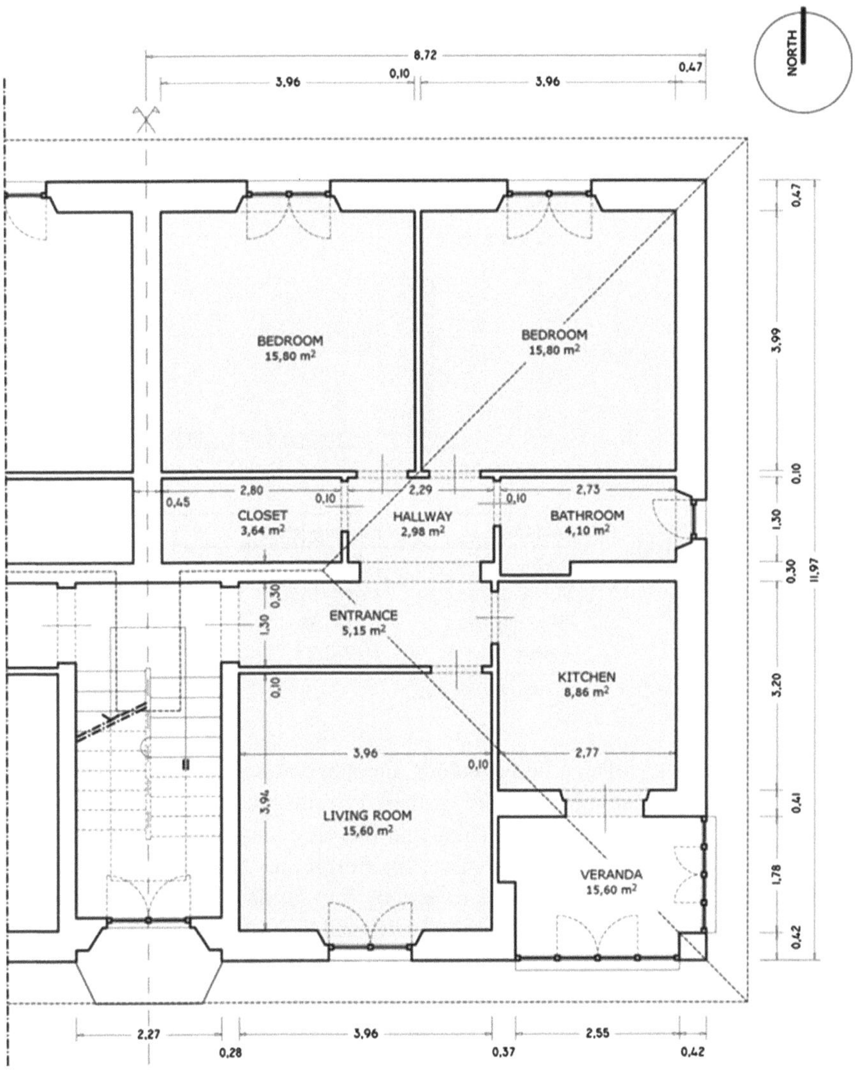

Fig. 17 Plan of typical floor analyzed

Once defined the calculation codes more suitable for the objectives of the project, the knowledge of the environment peculiarities of the building site becomes fundamental.

In the following, the reference to four locations is considered: Berlin, Milan, Florence, and Athens (Table 1).

As for the energy simulations, for each weather zone, the dry bulb temperature of the outside air (θ_{db}) and the solar radiation (I_{sol} expressed in W/m^2) over the various orientations are required to evaluate thermal loads during winter and summer times.

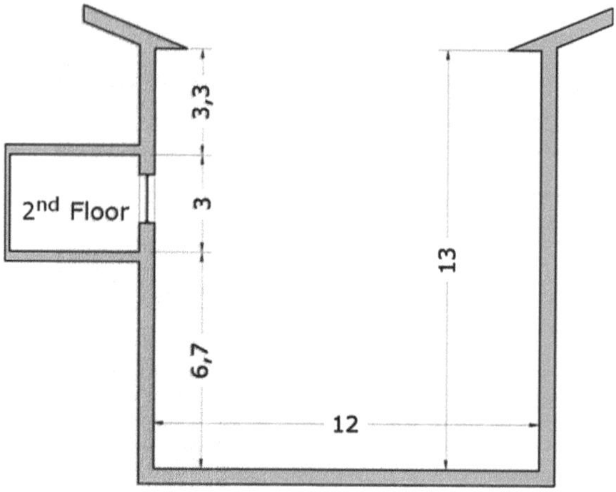

Fig. 18 Hypothesis of the urban context (section)

Table 1 Locations considered in the analyses

Location	Heating period	Cooling period
Berlin	1/10–30/4	1/6–31/8
Milan	15/10–15/4	15/5–30/9
Florence	1/11–15/4	1/5–15/10
Athens	1/12–15/3	1/5–15/10

The hourly values of dry bulb outdoor air temperature and solar radiation to perform the energy simulations are gathered from the Institute “Gianni De Giorgio” (IGDG) [31] archive, for Milan and Florence, and from the International Weather for Energy Calculations archive, for Berlin and Athens.

The case study shows dimensions of 4 m by 4 m in plant and 3 m in height of the ceilings. In Fig. 19, the geometrical features and the position of the window are shown and in Fig. 20 the case with the French door is presented.

It is assumed that the room is located on the second floor (height from the road level equal to 6.7 m) of a 4-storey building, 13 m high. The cell type used for the calculations is reported in Figs. 19 and 20.

For the specific purposes of the lighting evaluations, the light reflection coefficient of inner surfaces has been assumed equal to 0.6 for the walls and ceiling (plaster and furniture in clear color) and 0.4 for the floor.

The light reflection average coefficient of the external surfaces and of the front building façade was assumed to be equal to 0.6, and then, three walls of the cell and the two horizontal partitions are considered adiabatic. The window is located in the fourth wall, which presents thermal losses depending on the orientation. Therefore, different case studies are evaluated (North, South, West and East).

The basic configuration of the cell presents a mixed masonry external wall, plastered on both sides, of 0.47 m total thickness. The window is single-glazed with 3-mm-thick panes, and the wooden frame is 50 mm thick, corresponding to

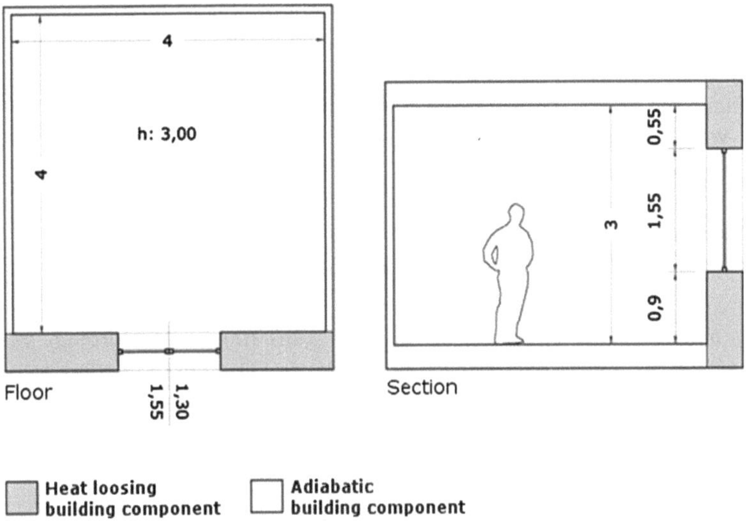

Fig. 19 Plan and section of the case study

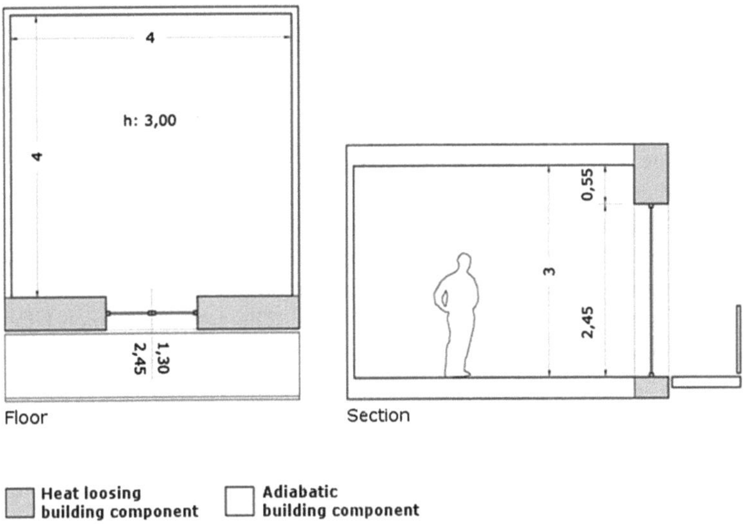

Fig. 20 Plan and section of the case study with French doors onto a balcony

20 % of the surface of the window frame. Tables 2 and 3 show the main thermal and acoustic performances of the external wall and of the window.

Appropriate performance indicators (defined by regulations or conventionally applied) should be identified, in order to evaluate different strategies for the achievement of energy efficiency and of comfort requirements. The same indicators can be applied to compare energy consumption, lighting, and acoustic improvement solutions suggested through the analysis procedures.

In the current case study, the following performance indicators were identified, to evaluate the energy performance of different refurbishment strategies:

Qs_w Winter solar gains (kWh);
Qs_s Summer solar gains (kWh);
θ_o Operative temperature (°C);
F_w Reduction factor of winter solar gains (%);
F_s Reduction factor of summer solar gains (%).

Qs_w and Qs_s represent the solar thermal gains transmitted through the glass panes of area A_g, evaluated during heating (W) and cooling (S) period, in relation to the incident solar radiation I_{sol}, from the glass area A_g.

Seasonal reduction factors (F_w and F_s) are calculated as the complement to the unity of the ratio of the solar gains transmitted through the glass on an hourly basis, respectively, in winter Qs_w or in summer Qs_s to those of the reference case.

These indicators can be used to compare the solar gain reduction effectiveness over a building with or without the adoption of solar shading devices.

To evaluate the performances of different lighting strategies, the following parameters were applied:

DF Average daylight factor (with standard overcast sky) (%);
UDI Useful daylight illuminance (–);
E_{min}/E_m Daylight uniformity (with standard clear sky) (–).

The average DF indicates the percentage of natural light in the indoor environment in overcast conditions.

The UDI is referred to annual time series of absolute values for illuminance predicted under realistic skies generated from standard meteorological datasets. It expresses the annual occurrence of illuminances on the work plane, where all the illuminances are within the range 100–2,000 lux. The degree to which UDI is not achieved because illuminances exceed the upper limit is indicative of the potential for occupants' discomfort [13, 14].

The uniformity of natural light (E_{min}/E_m) provides information on the inner distribution of lighting, with clear sky. It represents an important factor, because a poor distribution of natural light leads to increase the need for artificial light and thus energy consumption for lighting.

The "clear sky" model corresponds to that described by the International Commission on Lighting (Commission Internationale de l'Eclairage—CIE) [18, 19], which provides a luminance distribution as a function of the sun position at a given latitude.

The acoustic response of the vertical envelope composition has been weighted by determining the following parameters:

$D_{2m,nTw}$ Standardized façade sound level difference (dB);
ΔL_{fs} Façade shape level difference (dB).

Table 2 Performances of the opaque external wall

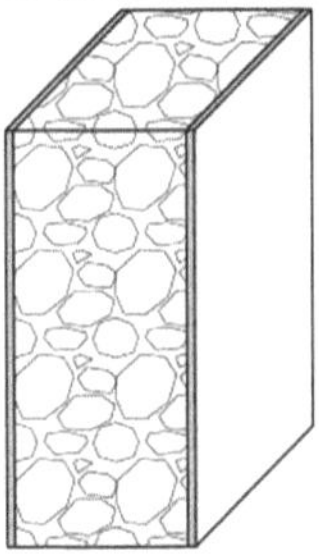	*Thermal performance*	
	Transmittance U	1.45 W/(m^2 K)
	Periodic thermal transmittance Y_{IE}	0.152 W/(m^2 K)
	Phase shift φ	14.19 h
	Acoustic performance	
Lime plaster and cement $s = 2$ cm	Sound reduction index R_w	56 dB
Masonry clay bricks and rubble $s = 43$ cm	Noise absorption coefficient of the external surface (63–4,000 Hz)	0.05–0.04–0.02 0.04–0.05–0.05

Table 3 Performances of the window

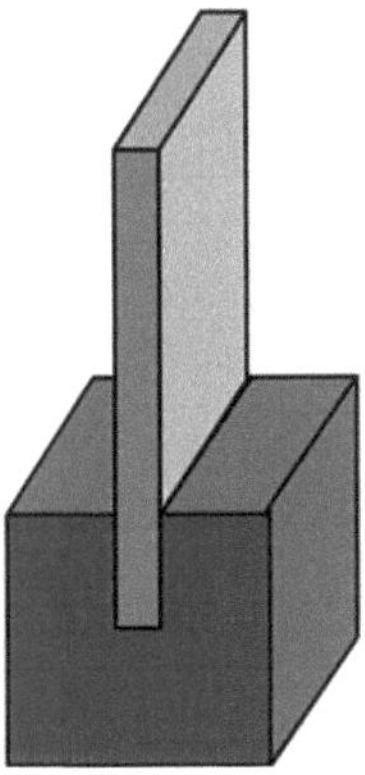	*Thermal performance*	
	Glass transmittance U_g	5.8 W/(m^2 K)
	Frame transmittance U_f	2.4 W/(m^2 K)
	Window transmittance U_W	5.33 W/(m^2 K)
	Glass solar factor g	0.87
	Acoustic performance	
	Sound reduction index R_w	28 dB (EN 12354-3)
Wood Frame $s = 5$ cm	Air tightness class	1 (EN 12207)
Clear glass single pane $s = 3$ mm	*Lighting performance*	
	Glass light transmission factor τ_v	0.80

The sound insulation of the façade has been determined from its shape and surface, the performance and surface of single components, and the type and quality of the sealing of the joints, as specified in EN ISO 12354-3:2000. The façade shape level difference has been simulated using the technique of ray tracing, taking into account the following variables: geometry of the system, absorption of façade components, and sound spectrum of the source.

Defining a consistent evaluation methodology in terms of energy, lighting, and sound efficiency of design strategies is crucial.

The analysis methodology hypothesized in this study refers to the most frequent sequence of the building energy refurbishment that can be found in several

Table 4 Summary of the simulations phases performed in this study

Phase A	Phase B	Phase C	Phase D
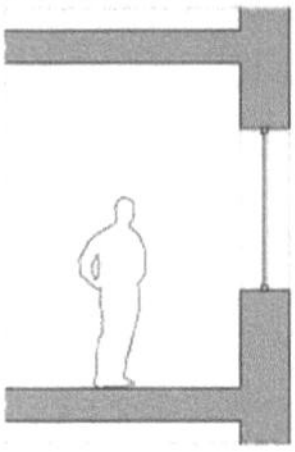	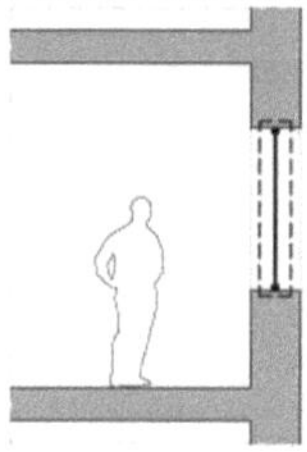	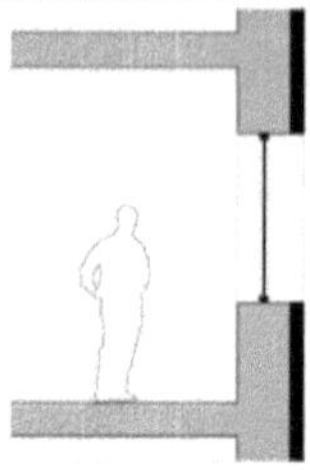	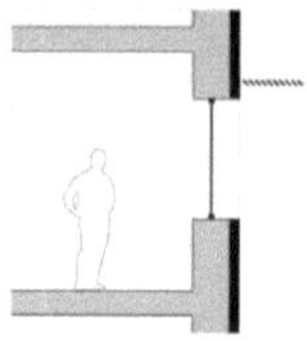
Existing building	Replacement of existing windows	Replacement of existing windows and improvement in the thermal transmittance of opaque envelope	Replacement of existing windows, improvement in the thermal transmittance of opaque envelope, and introduction of solar control systems

practical cases (Table 4): in existing buildings, the replacement of windows (49 %) is followed by the improvement in the energy performance of opaque vertical envelope (30 %).

Starting from the performance evaluation of an existing building (phase A), the study provides for the assessment of the following interventions (phases B–D), according to a logic consequential implementation performance:

- replacement of existing window with a high-energy performance one (phase B);
- adaptation of the thermal transmittance of opaque envelope to national legislation limits (phase C);
- introduction of solar control systems (phase D).

In the phase D, the configuration corresponding to the phase C is associated with different screening systems. As an alternative to screening systems, the performance of two different solar control glasses, which, respectively, present $g = 0.46$ and $\tau_v = 0.58$ and $g = 0.21$ and $\tau_v = 0.40$, has been evaluated.

The selection criteria for shading systems (Table 5) follow the requirements for the reduction in energy consumption in summer and in winter and the achievement of the best possible comfort keeping the view of the external environment in all seasons [15]. Among the possible materials available on the market, for weight and installation advantages, a metal product (aluminum) has been chosen; its white shining color reflects incident solar radiation, both direct and diffuse, with hemispherical uniform distribution.

Specifically, about the coating treatment and the overall performance the reference are the EN 14351-1, EN 15193, EN 10077-1, and the EN 410 standards.

Table 5 Schedule of different types of shielding analyzed in phase D

	D1.1—Sunshade with continuous horizontal blinds (tilting: 30°, 60°, 90°) Blind dimensions: 0.08 m × 0.90 m Step between the blinds: 0.08 m Overall width: 1.90 m Overhang: 0.8 m Distance from the window: 0.2 m
	D1.2—Venetian blinds (tilting: 0°, 30°, 60°) Blind dimensions: 0.08 m × 1.3 m Step between the blinds: 0.08 m Distance from window: 0.1 m
	D1.3—Sunshade with continuous horizontal and vertical screens Blade dimensions: 0.2 m × 1.5 m Step between the blades: 0.2 m Overhang: 0.8 m Distance from the façade: 0.8 m Distance from window: 0.8 m
	D1.4—Overhang opaque fixed horizontal Overhang: 0.8 m Width: 1.9 m Distance from the window: 0.2 m
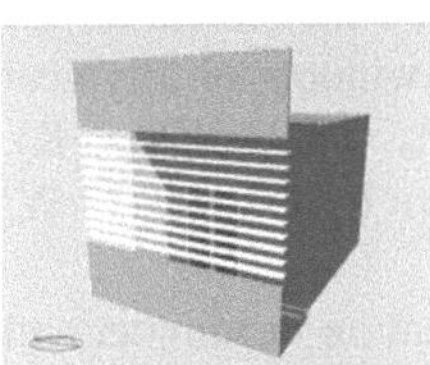	D1.5—Vertical sunshade with blades tilted integrated with balcony (tilting: 0°, 30°, 60°) Blind dimensions: 0.2 m length × façade Step between the blades: 0.2 m Balcony depth: 1.2 m
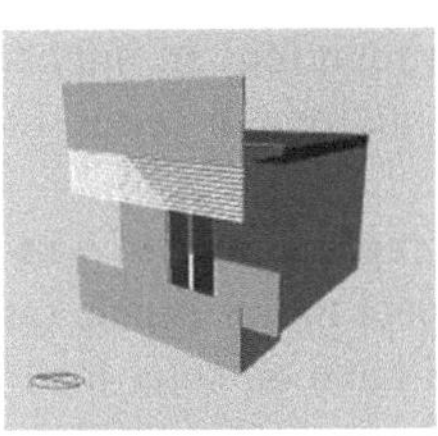	D1.6—Vertical sunshade with blinds tilted integrated with balcony (tilting: 0°, 30°, 60°) Blind dimensions: 0.08 m length × façade Step between the blinds: 0.08 m Shielding height: 0.8 m Balcony depth: 1.2 m

(continued)

Table 5 (continued)

	D2.1—Solar control glass $g = 0.46$ $\tau_v = 0.58$ $U_g = 1.60$ W/(m^2 K) $U_w = 1.77$ W/(m^2 K) D2.2—Solar control glass $g = 0.21$ $\tau_v = 0.4$ $U_g = 1.60$ W/(m^2 K) $U_w = 1.77$ W/(m^2 K)

6.2 Effect of Screens and Windows on the Summer Thermal Loads and Thermal Comfort

The presence of glass surfaces ensures winter favorable thermal gains; however, in summer, it may cause interior overheating by the sun.

Recently, the replacement of windows in existing buildings has become common practice, thanks to tax incentives offered in certain European countries, to the ease of implementation, and to the synergy of positive effects that the intervention may produce (for example, the improvement in the acoustic performance of the façade). This action presents a great deal of technical feasibility (TF) since it rarely involves outside interventions, for example, with scaffolding, and does not interfere seriously with the activities inside rooms.

Divided by location and window orientation, Fig. 21 shows the effects on the reduction in solar gains resulting from the replacement of a window of a typical building of the second post-war period (phase A, $U_w = 5.33$ W/(m^2 K), $g = 0.87$), with a high-energy performance window (phase B, $U_w = 1.77$ W/(m^2 K), $g = 0.58$).

The figure highlights the critical regime of the West orientation in summer and the need for a conscious choice of solar shading system to control the solar radiation loads, without penalizing winter solar gains, which represent an important contribution, especially with regard to the South orientation, in terms of energy savings.

An additional factor to be taken into account during the replacement of windows is the influence that the position of the frame with respect to the façade (at the outer edge, on the center line, and at the inner edge) can have on solar loads and consequently on the need for air-conditioning.

For instance, for the climatic conditions of Florence, as shown in Fig. 22, the position of the window with respect to the façade involves, in the transition from the inner to the outer edge, an increase in winter solar gains between 48 % (North

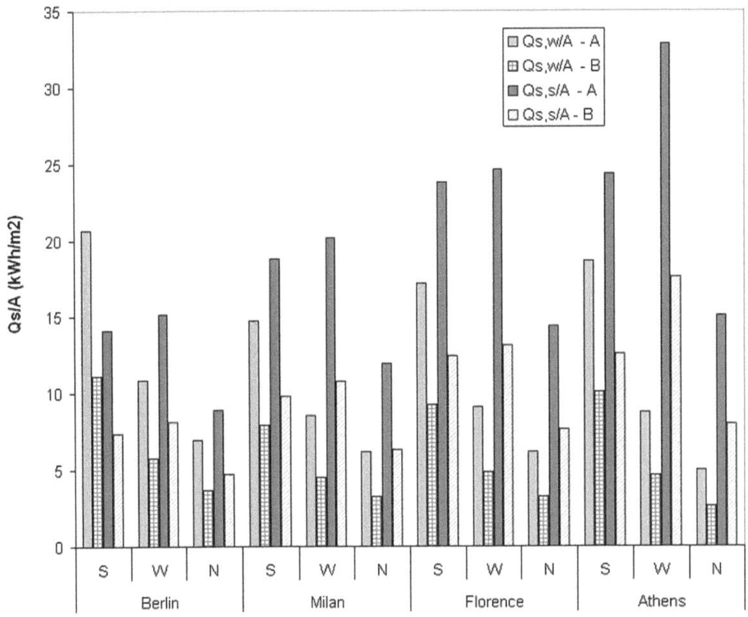

Fig. 21 Winter (Qs_w) and summer (Qs_s) solar gains per unit floor area, selected by location and window orientation relatively to the phases A and B

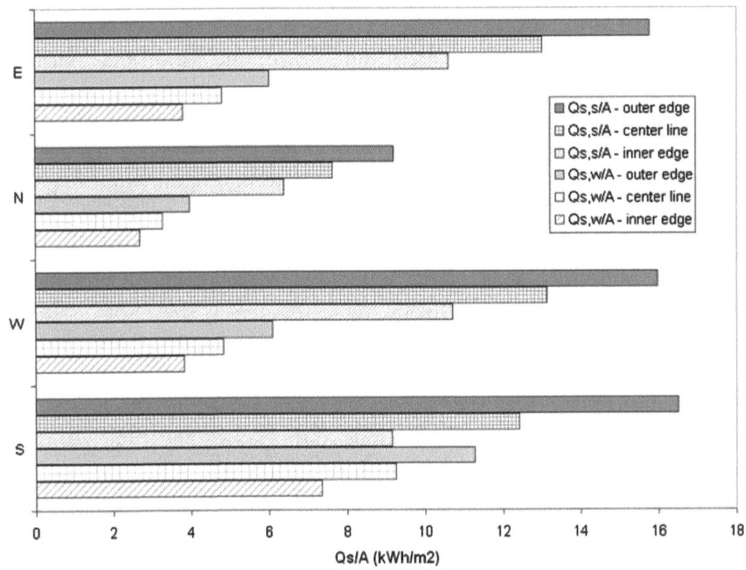

Fig. 22 Winter (Qs_w) and summer (Qs_s) solar gains per unit of floor area for the climatic conditions of Florence, selected by orientation and position of the window with respect to the wire of the façade

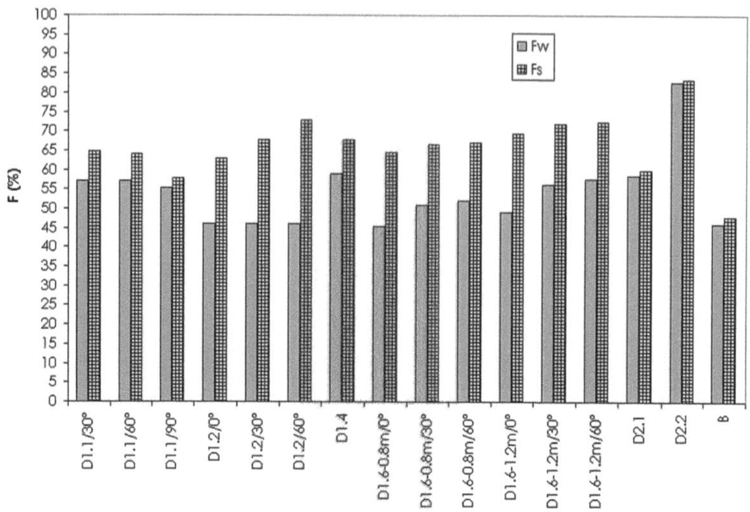

Fig. 23 Seasonal reduction factors (F_w and F_s) for different shading systems selected for Milan, South orientation

orientation) and 59 % (West orientation) and an increase in summer solar gains between 44 % (North orientation) and 80 % (South orientation).

The different position of the frame, relatively to the façade profile, leads to the resolution of some technological details, which concerns mainly the relation with the thermal insulation, the reduction in thermal bridges, the presence or the absence of space where to place the shading system, and the proper sealing on the frame/masonry coupling in order to prevent infiltration of air and noise.

The legislative developments of recent years have given great importance to the need for solar radiation control in summer conditions, forcing the designer to consider the problem of verifying the risk of indoor overheating due to unshielded glass surfaces.

In the absence of effective solar radiation shielding, some refurbishment actions on the existing buildings, such as the reduction in the opaque envelope thermal transmittance, may even increase, rather than decrease, the need for air-conditioning systems in summer and get worse conditions of indoor thermal comfort.

The use of different types of solar shading or glass with solar control can improve the performance of the system window—shading [1, 6, 8], ensuring adequate indoor comfort conditions in both summer and winter and reduction in energy consumption for climate control of the building.

Since shading systems usually are not placed on the North façade, the results from dynamic simulations are compared for South and West façades (the most critical in the summer), resulting from the application of the solar radiation control systems most commonly used in residential buildings, excluding intrusive or hardly feasible configurations.

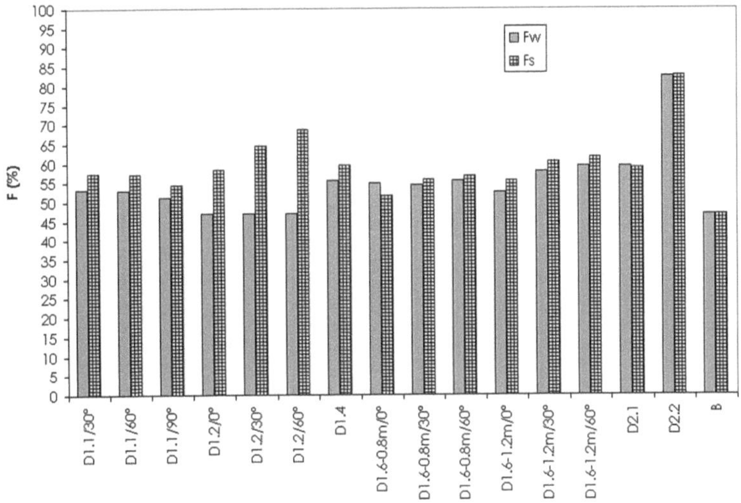

Fig. 24 Seasonal reduction factors (F_w and F_s) for different shading systems selected for Milan, West orientation

The effect of reduction on solar gains was evaluated through the seasonal reduction factor, respectively, in winter (F_w) or in summer (F_s), expressed as a percentage.

In Figs. 23, 24, 25, 26, 27, and 28, relative to the location of analysis, the seasonal reduction factors derived from the comparison with the existing building (phase A) are shown referred to South and West exposure in the various locations; in particular, the D1.2 shading system was considered packaged in the winter season, in agreement with the most common use.

In general, among the several relevant parameters for the choice of a control system of the solar radiation, the capability of the shading system to reduce thermal loads in summer and at the same time to allow solar gains in winter must be taken into account.

This feature can be analyzed by comparing the difference between F_s and F_w (ΔF) extrapolated from Figs. 23, 24, 25, 26, 27, and 28; in substance, a shading system may be considered much more effective if it has a high value of F_s and a corresponding low value of F_w, and then, higher ΔF value corresponds to greater shading effectiveness.

For all the analyzed locations, representative of the different European climate conditions, the values of ΔF were evaluated for the South facing.

The results highlight that the Venetian blind (D1.2) is the most efficient system, when it is considered completely packed in winter; in particular, the blinds tilted to 60° are the most effective. To ensure these benefits, combining this shading system with a building automation system that manages the opening in a dynamic way might be useful.

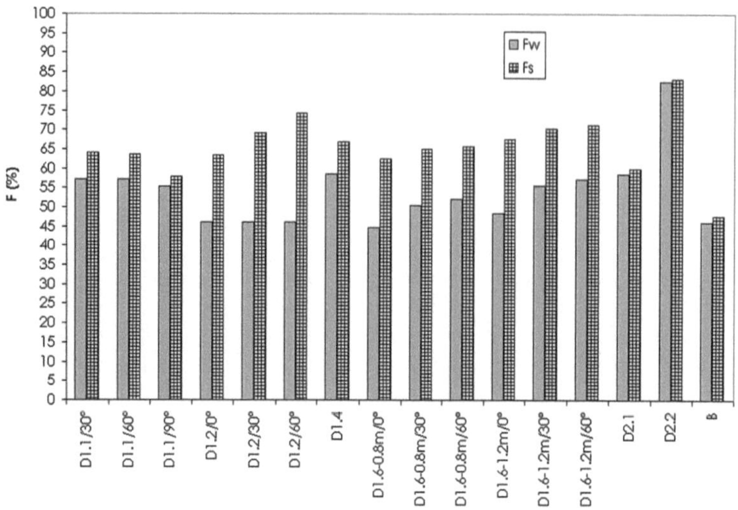

Fig. 25 Seasonal reduction factors (F_w and F_s) for different shading systems selected for Florence, South orientation

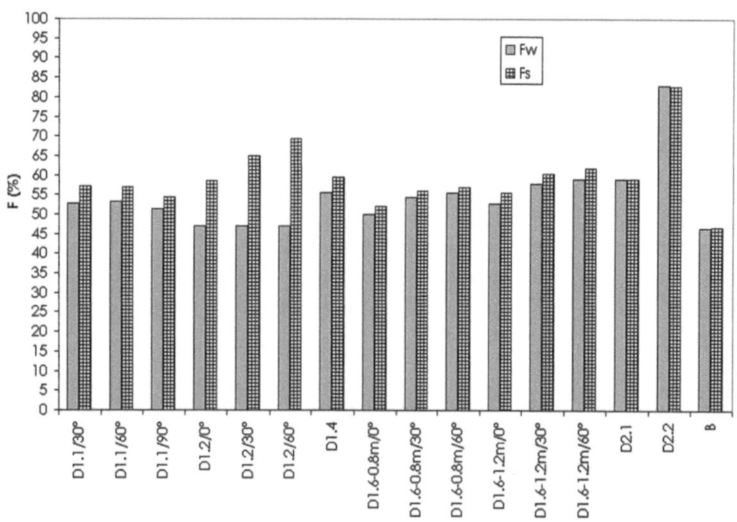

Fig. 26 Seasonal reduction factors (F_w and F_s) for different shading systems selected for Florence, West orientation

The insertion of a shading system (D1.6) in an existing balcony or the addition of a new balcony adjacent to the existing building gives good results, especially when it is combined with horizontal blinds tilted to 0°. In this case, the depth of the balcony (analyzed between the dimensions of 0.8 and 1.2 m) affects in a limited manner.

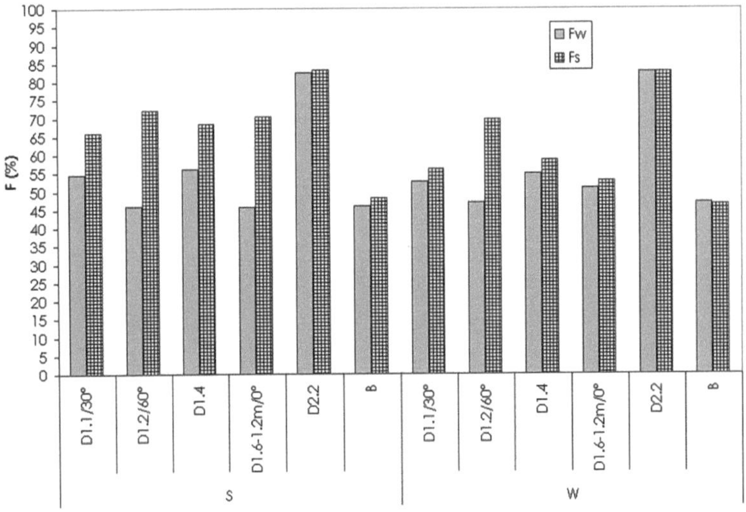

Fig. 27 Seasonal reduction factors (F_w and F_s) for different shading systems selected for Berlin, South and West orientations

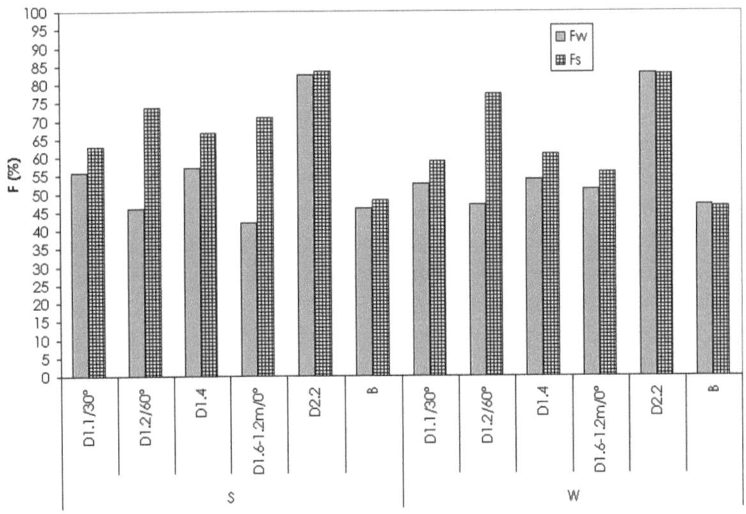

Fig. 28 Seasonal reduction factors (F_w and F_s) for different shading systems selected for Athens, South and West orientations

The shading system perpendicular to the façade with horizontal blinds (D1.1) and the opaque horizontal overhang (D1.4) have very similar performance and far lower than previous analyzed (on the order of 50 %). In particular, the D1.1 typology achieves the best performance for blinds tilted to 30°.

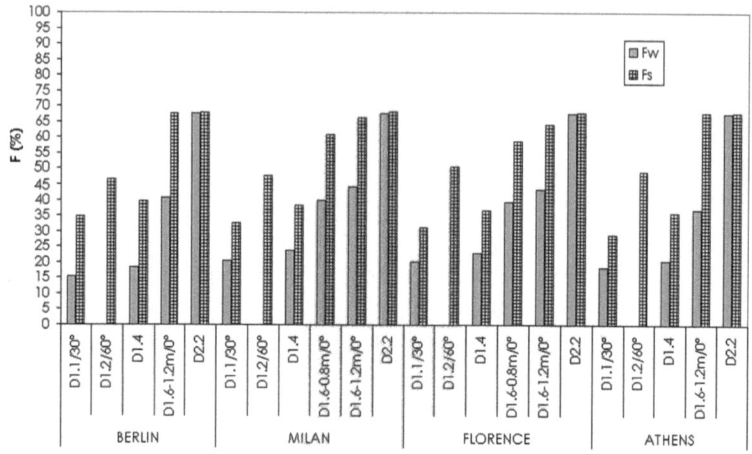

Fig. 29 Seasonal reduction factors (F_w and F_s) for different shading systems, relatively to South orientation

The solar control glasses have the same solar gain reduction, both in summer and in winter, so their use should also be evaluated as a function of the intended use of the property. This strategy can be considered a valuable alternative to the use of external shielding in situations in which the insertion in the façade of extraneous elements to the original morphology of the building is problematic (such as, for example, in the case of historical buildings and in historical centers) or technically complex.

Regarding the West-facing position, the most effective shading system would have vertical blades or blinds, which, however, are rarely used in residential applications. Among the analyzed sunshades, which have in general a lack of effectiveness for this orientation, the external Venetian blind (D1.2) with an inclination of 60° is the most performing.

In Fig. 29, related to South exposure, relative to the locations of analysis, the seasonal reduction factors of the main sunshades that are reported are compared with the replacement of the window (phase B).

This comparison is useful when the designer has already started a process of energy retrofit of the building envelope. This phase can then be seen as a further implementation of the performance, in order to contain energy consumption and improve indoor comfort in summer conditions.

The arising considerations confirm the effectiveness of the analyzed screens for South-facing position and emphasize the highly efficient behavior of the external Venetian (D1.2), the configuration with balcony and integrated shield (D1.6), the sunshade perpendicular to the façade with horizontal blinds (D1.1), and the horizontal overhang (D1.4).

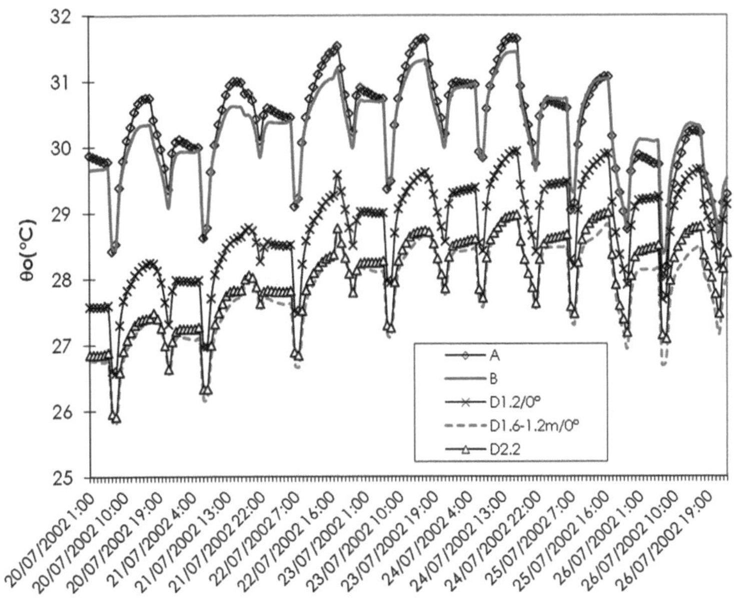

Fig. 30 Operative temperature trend within the South-facing cell located in Florence for different sunshade systems in a typical summer week

It is apparent that the seasonal reduction factors cannot be the only parameters that influence the shading design process, since they do not take into consideration a number of fundamental questions, such as user's comfort, cost, TF of the intervention, and the morphological integration with the building, which will be discussed later.

In particular, the thermal comfort of the occupants can also be estimated by means of the operative temperature. In order to assess, although in a preliminary manner, the implications on the thermal comfort of some solar control systems, Fig. 30 shows the trend of the operative temperature inside the cell type exposed to the South, represented in a summer week (July 20–26) for the location of Florence, while Fig. 31 shows details related to the 23 and 24 of July.

The values for the following configurations are compared: existing building (phase A), replacing windows (phase B), insertion of different sunshade, composed by external Venetian blinds with an angle of 0° (phase D1.2/0°), shielding system integrated on the 1.2-m balcony and blinds with an angle of 0° (phase D1.6–1.2 m/0°), and solar control glass with $g = 0.21$ (phase D2.2).

The simple replacement of the window reduces by little the operative temperature, while the application of screening systems produces a reduction in the operative temperature ranging from 2 °C (with Venetian blinds) to about 3 °C (with integrated system on the balcony).

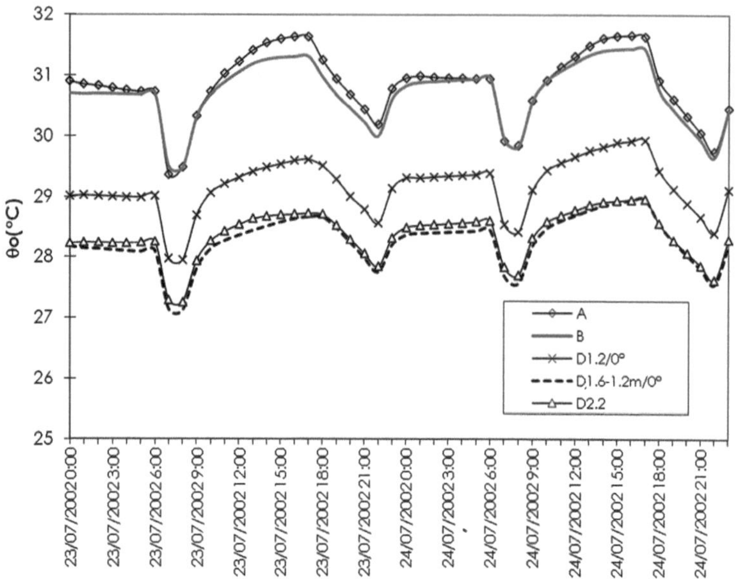

Fig. 31 Detail of the operative temperature trend within the South-facing cell located in Florence for different sunshade systems

The use of solar control glass leads to operative temperature values comparable with those of the integrated balcony systems, proving its effectiveness in summer time, subject to the risk of penalization during winter.

In general, the use of sunshade systems as passive control techniques of the indoor conditions involves both an improved comfort and a reduction in the air-conditioning need in summer season.

6.3 The Shading Effect on Visual and Acoustic Comfort

The windows and the shielding system performances may change significantly both the distribution of daylight, and the thermal and acoustic comfort. Below, the influence on visual and acoustic comfort of different configurations of windows and shielding is described.

6.3.1 The Influence of Shadings on Daylight Distribution and Visual Comfort

The natural light simulations are referred to the systems A, B, C (B and C are equal for this purpose) and D, with the calculation assumptions specified in Sect. 6.1 and using the software RELUX.

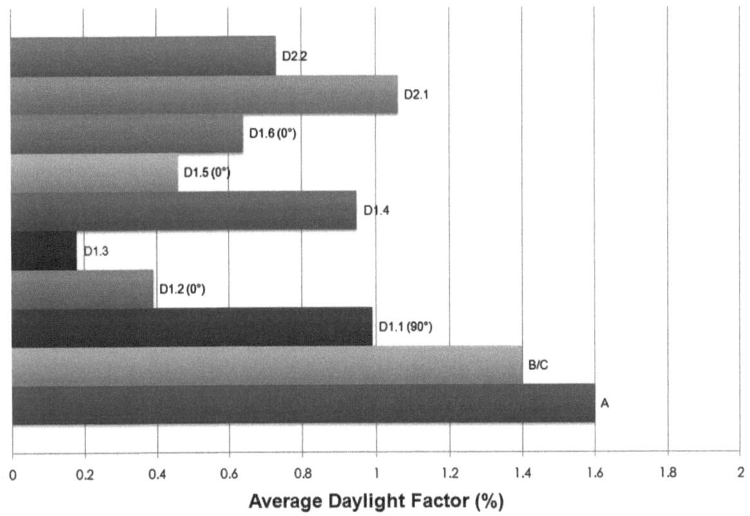

Fig. 32 Average daylight factor with standard overcast sky (*numbers in parentheses* indicate the tilt of the slats)

The influence of different shielding systems and of the kind of glass on the quantity and quality of daylighting has been assessed with reference to different orientations of the façade. The North orientation has been omitted, because it is assumed that on this side, there are no shielding systems. Results referring to East and West are averaged because of their little difference.

The performance evaluation of the different systems is based on the following parameters, already described in Sect. 6.1:

DF Average daylight factor (with standard overcast sky) (%);
UDI Useful daylight illuminance (–);
E_{min}/E_m Daylight uniformity (with standard clear sky) (–).

The results related to shielding systems D1.5 and D1.6 are referred to a balcony 1.2 m deep.

Figure 32 shows the average DF values for different shielding systems. The analyzed shielding, under overcast sky conditions, significantly reduces the level of natural lighting inside the examined room.

For South exposure, nevertheless they guarantee the maintenance of a good level of natural lighting, as shown in the graph of UDI (Fig. 33).

Almost all the examined shieldings provide illumination levels between 100 and 2,000 lux, more than 80 % of the time during the year, for Southern exposure (Fig. 33). With West or East exposure, shielding types D1.2 and D1.5 guarantee the requirement for 50 % of the time. The remaining time, UDI is less than 100 lux. These shields, if not adjustable, are therefore excessively unfavorable for exposures other than that of South.

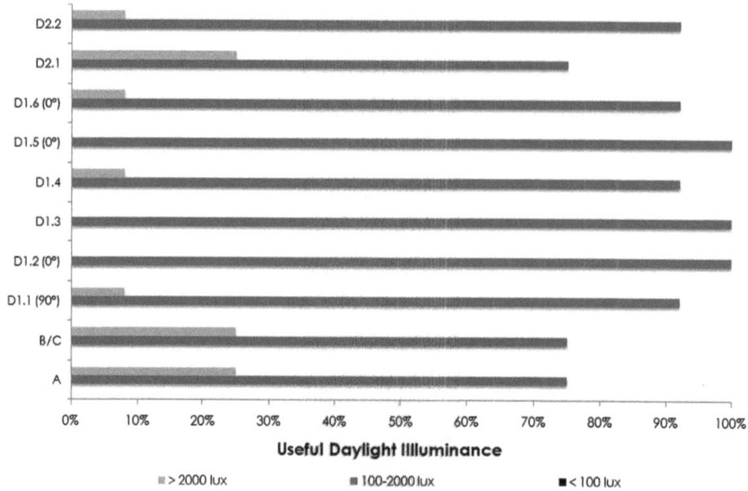

Fig. 33 South façade useful daylight illuminance (*UDI*)

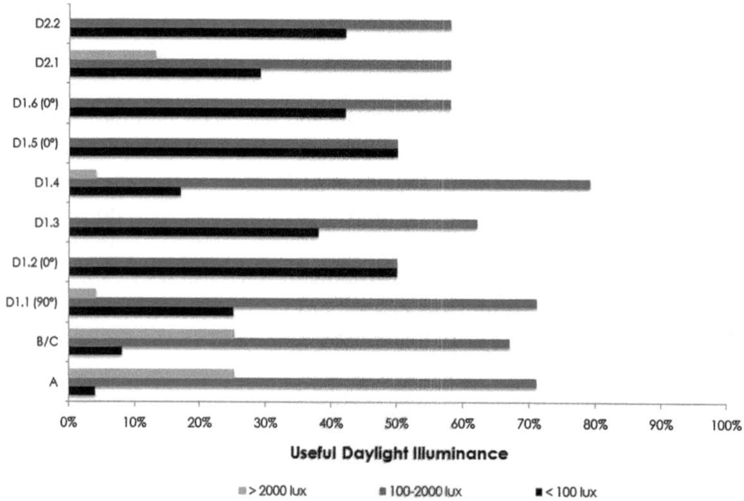

Fig. 34 East and West façades useful daylight illuminance

Figures 32, 33, and 34 show that the systems with inclined blades (D1.2 and D1.3) cause an excessive reduction in the natural lighting level with overcast skies. Therefore, these screens should always be equipped with a mechanism for adjusting the slat inclination. However, with clear skies, they allow a reasonable level of daylighting especially for South-exposed façades, even with fixed and inclined slats.

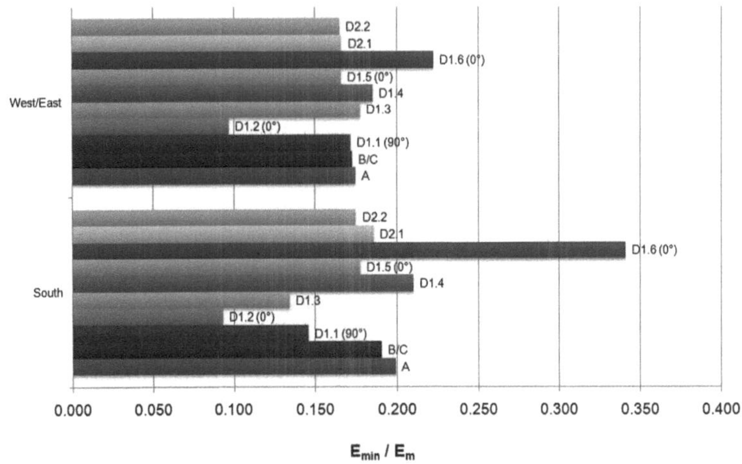

Fig. 35 Uniformity of natural light estimated in clear sky conditions

The shielding system D1.6 represents a good compromise for Visual comfort as it ensures a natural lighting level with clear sky sufficient for most visual tasks for more than 90 % of the time for South exposure and about 60 % of the time for East or West exposure. In addition, the distribution of natural light with clear sky is significantly improved.

The solar control glasses (D2.1 and D2.2), when exposed to South (Fig. 33), are not always appropriate because they can determine internal lighting values that produce visual or thermal discomfort at certain times of the day. For the East or West (Fig. 34) exposure, in certain periods, the natural light must be integrated with the artificial to have a sufficient internal lighting level.

6.3.2 Improvement in Façade Acoustic Performance Due to Shielding Systems

The sunscreens, if well designed, can work as acoustic screens, therefore improving the performance of the façade, ensure significant noise protection of the interior, even with the open windows.

For this purpose, it is necessary that the size and the inclination of the blades of the sunscreens are suitable to intercept all the sound waves coming from the external sources.

In the case of buildings faced to streets, these sources are usually represented by the traffic and therefore are placed at the street level. In these conditions, the horizontal arrangement of the slats reflects the sound waves before reaching the plane of the façade. Therefore, if the lower surface of the blades is coated or made with highly sound-absorbing material, the sound waves are heavily attenuated during reflection, before arriving to the façade (Fig. 36).

Fig. 36 Effect of a sunscreen of a façade on the sound waves coming from the street. On the right, the detail of the sound level attenuation

In addition, to further increase the effect of sound attenuation, even the surface of the balconies that looks down should be coated with highly absorbing materials.

In order to evaluate the screening effect on the sound pressure reduction in front of buildings, the following parameters are considered, as defined in the standard EN 12354-3 [24]:

$D_{2m,n\mathrm{Tw}}$ Standardized façade sound level difference (dB);
ΔL_{fs} Façade shape level difference (dB).

The acoustic simulations, performed with the software DISIA [11], allow to analyze the acoustic performance of the upper floors façade. The ground floor, in fact, is merely influenced by the screen effect, because the sound waves coming from the road are directed perpendicular to the façade and the shielding effect of the window sills and of screening system becomes therefore negligible.

For the purposes of the simulations, the sound source, which represents the sound spectrum of the urban road traffic, was placed at the center of the roadway in front of the examined façade.

The effect of shielding types D1.1, D1.2, D1.3, D1.4, D2.1, and D2.2 is negligible on the acoustic insulation of the façade, because in these cases, the sound waves cannot be effectively intercepted and absorbed before reaching the façade.

Therefore, the results are referred only to the shielding types D1.5 and D1.6, with different depths of the balcony.

In Fig. 37, the values of ΔL_{fs} at different floors are presented. Numbers after the code of the system indicate the depth of the balcony.

Figure 38 shows the $D_{2m,n\mathrm{Tw}}$ values obtained at different floor levels by applying the proposed methodology. The types A and B, which refer to the façade without screening systems, with basic window (window $R_w = 26$ dB) and upgraded window (window $R_w = 32$ dB), are shown for comparison.

Furthermore, the results also take account of the fact that the application of shielding systems D1.5 and D1.6 involves the replacement of the window with a window door and therefore of the difference in size (1.95 vs. 2.75 m^2).

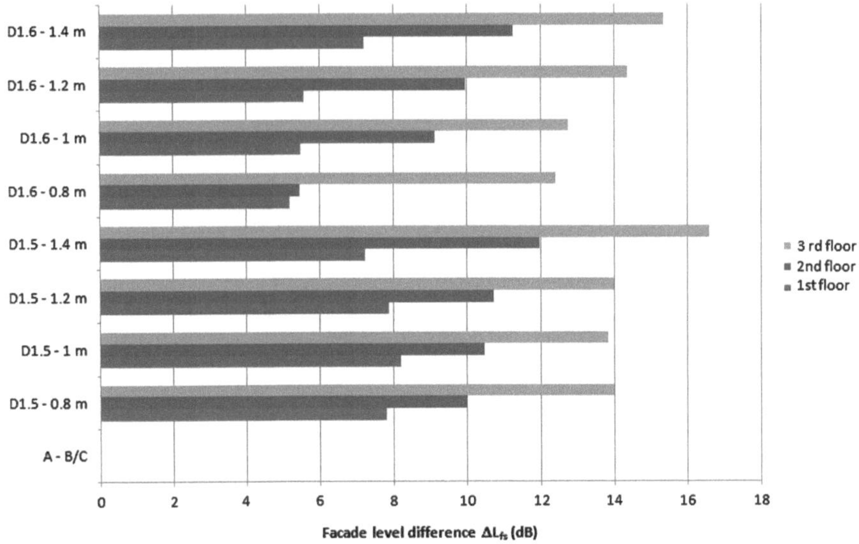

Fig. 37 Sound pressure level differences based on façade shape (ΔL_{fs})

The values referred to the first, second, and third floors are due to the different effectiveness of the shielding, as a consequence of the different angle of incidence of the sound waves, coming from road traffic. This effect is confirmed by the computational method values performed according to the EN 12354-3:2000 (variability in function of the height from the sound source) and by recent studies conducted by the authors [5].

The results demonstrate a strong increase in the sound insulation of the façade, especially at higher floor levels and with screening systems of type D1.5, with sound-absorbing materials built in terraces and louvres. This acoustic effect is particularly significant because it involves an improvement in acoustic comfort in the indoor environment, even in the condition of open windows.

7 How to Choose a Solar Shading Device

In this paragraph, a comprehensive evaluation of the aspects of energy consumption, natural lighting, acoustic comfort, and Technical Feasibility is carried out in the form of a summary.

In particular, the following main functional benefits are evaluated:

- solar gain reduction in summer;
- thermal winter solar gains;
- summer thermal comfort improvement by controlling the phenomena of radiative heat exchange;

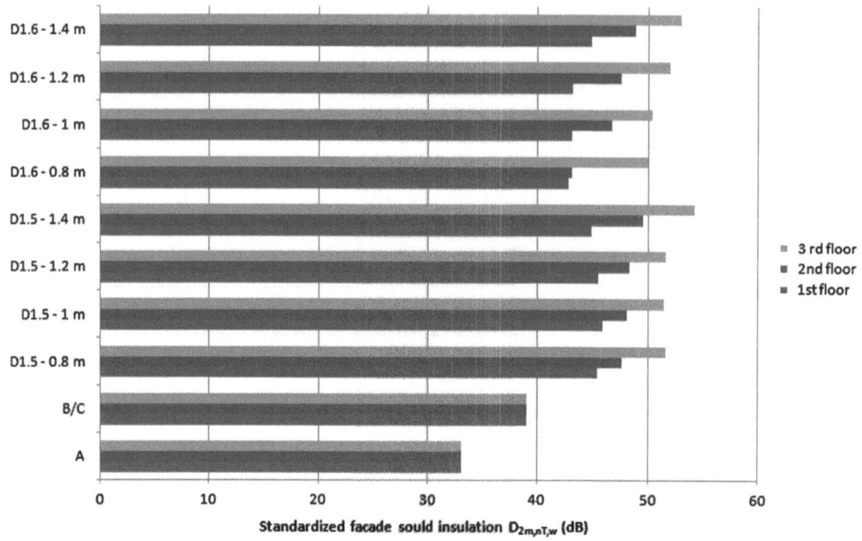

Fig. 38 Façade sound insulation ($D_{2m,nT,w}$)

- Visual Comfort by controlling glare effects, while maintaining the necessary contact with the outside perception in all seasons;
- acoustic comfort improvement;
- thermal resistance improvement in the combination frame/screen, when necessary.

7.1 Selection Criteria

The scheme of shielding typologies in relation to the achievable benefits, starting from the position of the shielding system, with respect to the window (external, internal, and intermediate), takes into account the main functional and performance benefits previously described (Table 6). The third type is widely used in office buildings and for double skin façades [15].

One of the targets of residential building refurbishment is to achieve the above-listed functional benefits; however, mainly due to structural and economic difficulties, external solar shadings are preferred, mainly in the areas with high levels of solar radiation (Mediterranean climate).

Table 7 shows the main types of solar shading systems for residential buildings, sorted according to their position relative to the window. Data are referred to current production, so variations are possible in terms of size and materials related to technological development in the sector.

Table 6 Functional benefits of solar shading with respect to its position

Position	Summer thermal gains	Winter thermal gains	Summer thermal comfort	Visual comfort	Acoustic comfort
External	++	−	++	+	+
Intermediate	+	+	+	+	−
Internal	−	++	−	+	−

Legend
++ Very favorable effect
+ Positive effect
− No effect or potentially negative effect

In Table 8, the different types of solar shadings described in the previous table are summarized, in order to provide preliminary guidelines on the most suitable types for each specific case [15].

In Table 9, the major aspects to be taken into account for the most appropriate screening system and the corresponding actions are reported.

7.2 Comparative Analysis

In order to choose correctly a solar shading system, a global comparative analysis has to be carried out.

The aim of this analysis is to define a method for the comprehensive evaluation of the shielding system, previously selected on the basis of the simulations carried out to assess their energy, acoustic, and lighting performance, besides Technical Feasibility and management problems.

This method is applied to the described case study and reported for typical sun-shading devices used in residential buildings.

These considerations are reported in the following data sheets (Table 10), consisting of the following sections:

- name of the system;
- technological details;
- analysis of the energy, daylighting, and acoustic behavior;
- synthetic solar shading evaluation.

The analyses of the energy, daylighting, and acoustic behavior were conducted for an unobstructed building, sited in Central Italy (Florence).

Results reported in data sheet concern the comparison between phase D (introduction of solar systems) and phases B/C (improvement in the envelope thermal performance).

The symbols in the data sheet express qualitative assessments (good, not relevant, or not satisfactory), associated with the screen typology. In particular, they express the relevance of the device in terms of the physical behavior response, with regard to the following requirements and performance indicators:

Table 7 Main types of solar shading systems applicable to residential buildings

Name	Description
External solar shading systems	
Horizontal sunscreen (example D1.1) 	The sunscreen consists of fixed horizontal blinds or grilles anchored to a structure perpendicular to the façade Blind material: extruded aluminum, bent or formed aluminum sheet, PVC-coated copper, wood, glass, PV panels, etc Structure material: aluminum, galvanized steel, etc Blade height (mm): 70–1,500 (with boring) Blade length (mm): max 6,000 Blind step (mm): 70–150
Fixed overhang (example D1.4) 	Overhang fixed horizontal, opaque, made with different materials (sheet metal, treated wood, plastic materials, PV panels, concrete, etc.). Anchored to the wall with an autonomous structure or structurally integrated. The shields may also have a vertical arrangement perpendicular to the façade; in this case, they are most effective for East and West orientations
Grating 	Overhang fixed opaque, made out of different materials, consisting of horizontal and vertical elements to create a grating pattern

(continued)

Table 7 (continued)

Name	Description
Sunscreen fixed blade (example D1.5) 	Outdoor solar shading preoriented blades fixed to the façade. This shading could also have vertical blades; this case, most effective for East/West orientation, is more frequent in commercial building. The blades can also be applied to shield balconies Blade section: ellipsoidal, arcuated, triangular, gull wing, etc Blade materials: extruded aluminum, formed aluminum sheet or bent, wood, PVC, brick, etc Horizontal blade height (mm): 25–1,200 Blade intersection (mm): 70–150 Max length (mm): 8,000
Venetian blinds (example D1.2 and D1.6) 	Solar shield for outdoor use with adjustable and packable blinds. The packaging of the blinds allows a very compact folded element once rolled in. The typology can also be applied to screen balconies other than windows Blind section: arched Blind materials: aluminum alloy, etc Blind supports: steel, etc Blind height (mm): 58–95 Blind width [mm]: 500–4,500 Screen height (mm): 400–5,000 Handling: hand winch crank, home automation systems for solar control
Persian shutter 	The opening of the shutter can be the classic hinged, folding, sliding. The blinds can also be adjustable, allowing good modulation of radiation and light They are applicable in residential buildings, suitable interventions in historical areas Blind material: wood, aluminum, PVC, etc Blind height (mm): 40–150

(continued)

Table 7 (continued)

Name	Description
Roller blind 	Sunscreen with mechanical roller blind. In some models, it is possible to obtain an adjustment of light and ventilation (due to the opening between the blinds). Moreover, it is possible to have the complete obscuration of the interior. It is also a safety guard Material slats: aluminum alloy prefinished, PVC, etc Material structure: aluminum, etc Roller blind height (mm): up to 3,000 Length slats (mm): 500–3,000
Roller curtain 	Sunscreen with mechanical roller curtain Fabrics: glass fiber, acrylic fiber, polyester, PVC, etc Material structure: aluminum, etc Curtain width (mm): up to 7,500 Curtain height (mm): up to 7,500
Sliding arm awning 	This is a combination of the roller curtain and a drop-arm awning, with the fabric dropping vertically and then projecting forward It allows the possibility of having a suitable shielding to solar radiation while allowing ventilation and visual contact with the exterior with lowered curtains
Tent canopy	This kind of sunscreen takes vantage of its convex-shaped canopy giving the possibility of a suitable shielding from solar radiation, while allowing ventilation and the vision of the exterior

(continued)

	A weatherproof casing permits the packing of the canopy Material structure: aluminum, etc Fabrics: opaque screen, waterproof polyacrylic, PVC, etc Overall width (m): 5 Overhang max (m): 2
Drop-arm awning 	Sunscreen with an arm projecting forward when lowered. It is equipped with a fabric and a head box for retracting the fabric It may apply to balconies, uncovered terraces, windows, etc Material structure: aluminum, etc Fabrics: opaque screen, waterproof polyacrylic, PVC, etc Overall width (m): 18 Overhang max (m): 5
Intermediate solar shading systems	
Venetian blind–Roller blind 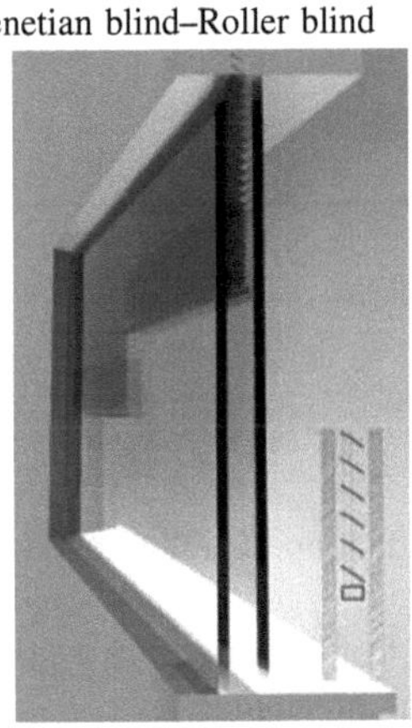	Double glazing which integrates into the interior chamber (of variable thickness) a Venetian blind, roller or pleated. The sliding of the tent takes place in a sealed package containing desiccants to ensure the control of humidity and vapor condensation Venetian blind, with respect to roller blind, provides a vision of the outside, even screening down, because it has oriented slats Applicable to windows of commercial and administrative buildings, schools, hospitals, and residences Max dimensions (mm): 32 (pleated and Venetian blinds) Handling: electrical, magnetic mechanism
Internal solar shading systems	
Vertical curtain	Solar shading mostly used to control daylighting, usually operable by hands over a rail system

(continued)

Panel width (mm): from 660 to 750 depending on the material
Rail material: aluminum

Venetian blinds

Solar shading device composed of slats of aluminum, wood, or plastic that adjusts by rotating from open to closed position by allowing slats to overlap
Mostly operated with cord or wand, also available in motorized version. Slats can be perforated
This solution is very common in commercial buildings, schools, hospitals, and residential buildings
Slat height (mm): 16–75 depending on the material
Width and height max (mm): 450

- Technical Feasibility (TF): installation, need for skilled manpower, and need for further building permits;
- Management (M): user's possibility to act on the effect of the shielding system, for instance by varying the angle of the blinds, easy maintenance, etc;
- Reduction factors F_w, F_s: reduction factors (respectively, for winter and summer and South orientation of the screen) of the solar thermal load, expressed as a percentage. In particular, when F_w is in the order of 20 % or less, the system is considered not affecting the solar gains;
- Visual comfort (VC): takes into account the uniformity of illumination and the amount of available natural light (UDI);
- $D_{2m,nT,w}$: rating of sound insulation of façade expressed in dB; in particular, it is considered "good" when the contribution of the system is at least greater than 1 dB.

Table 8 Evaluation of different types of solar shadings

Type of solar shading		Summer thermal gains	Winter thermal gains	Winter heat loss reduction	Summer thermal comfort	Visual comfort	Outdoor visual perception	Acoustic comfort	Wind resistance	Best orientation
External position with respect to the frame										
Fixed screens	Horizontal sunscreen	++	++	–	+	+	++	–	++	S
	Fixed overhang	++	++	–	+	+	++	–	++	S
	Grating	++	+	–	+	+	+	–	++	S–E–W
	Horizontal sunscreen fixed blades	++	0	–	++	0	0	–	++	S
	Vertical sunscreen fixed blades	+	0	–	+	0	0	0	++	E–W
Operable shielding	Venetian blinds	++	++	0	++	+	0	0	+	S–E–W
	Persian shutters	++	++	+	++	+	0	+	++	S–E–W
	Roller blinds	++	++	++	+	+	–	+	++	S–E–W
Curtains	Roller curtain	+	++	–	+	0	–	–	0	S–E–W
	Sliding arm awning	+	++	–	+	0	0	–	0	S–E–W
	Tent canopy	+	++	–	0	0	++	–	0	S–E–W
	Drop-arm awning	+	++	–	+	0	+	–	0	S
Position in the cavity of the double glazing and solar control glass										
Venetian blind	0	++	–	0	+	0	–	–	S–E–W	
Roller blind	0	++	–	0	0	–	–	–	S–E–W	
Solar control glass	++	–	–	++	0	++	–	–	S-E-W	

(continued)

Table 8 (continued)

Type of solar shading		Summer thermal gains	Winter thermal gains	Winter heat loss reduction	Summer thermal comfort	Visual comfort	Outdoor visual perception	Acoustic comfort	Wind resistance	Best orientation
Inner position with respect to the window frame										
Vertical curtain	–	++	–	–	0	–	–	–	–	
Venetian blind	–	++	–	–	+	0	–	–	–	

Legend
++ Excellent
+ Good
0 Moderate
– Not relevant
E East, *W* West, *S* South

Table 9 Main aspects to consider for a proper solar shading system choice

Aspects to take into account	Corresponding specific actions
Purposes of the intervention	Solar radiation and summer heat load control, glare reduction, aesthetic and functional rehabilitation, greenhouse effects' control, thermal and acoustic comfort improvement, etc
Historical buildings and landscape	A predetermined choice of shielding systems can be imposed, according to the historic features of the building and the site
Climatic location	Parameter collection and acquisition of the climatic conditions of the site (temperature, solar radiation, prevailing winds, etc.)
Window orientation	Seasonal variation evaluation of the incidence of solar radiation in relation to the environment (presence of shadows, boundary conditions, albedo effect, etc.)
Position on the façade	Position of the elevation of the screen in relation to solar energy and acoustic pressure, winds action, etc
Choice of the shielding system	Type (fixed or mobile), arrangement (horizontal or vertical), and tilt of the flaps, blinds, blades of the screen. Thermal comfort, light, and acoustic performances of the sunscreen
Technical feasibility	Building typology and compatibility with the chosen system: appropriate anchoring techniques, assembly and installation, etc
Management	Management and possibilities of operation, user friendliness, etc
Costs	Cost analysis, comprehensive of the installation
Costs of maintenance	Maintenance cost analysis, easiness to replace, availability of materials and spare parts, skilled manpower, etc
Costs/performances analysis	Costs/performance final evaluation, taking into account all the aspects above examined

Acknowledgments Authors thank Lorenzo Giorgi, Elisa Nannipieri and Leone Pierangioli for the valuable scientific collaboration in the drafting of this chapter.

Table 10 Data sheet for different solar shading devices—synthetic comparative evaluation

<table>
<tr>
<td rowspan="5">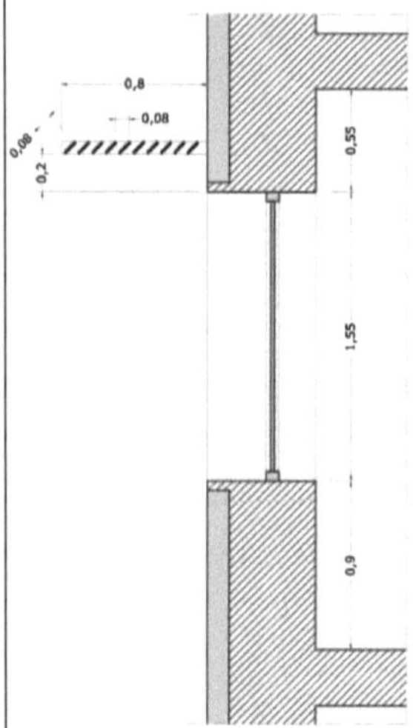
</td>
<td colspan="6">D1.1 - Sunshade with continuous horizontal blinds (tilting: 30°, 60°, 90°)
Best energy performances for south orientation and 30° blind tilting: Fs = 30% and Fw = 20%. Adequate daylighting. If west/east oriented, basic visual tasks are not guaranteed for the 30% of the year.
This shading does not affect the sound insulation of the façade.</td>
</tr>
<tr><td colspan="6">Synthetic Solar Shading Evaluation</td></tr>
<tr><td>TF</td><td>M</td><td>F_w</td><td>F_s</td><td>VC</td><td>$D_{2m,nT,w}$</td></tr>
<tr><td>☺</td><td>☹</td><td>😐</td><td>☺</td><td>☺</td><td>😐</td></tr>
<tr><td colspan="6"></td></tr>
<tr>
<td rowspan="4">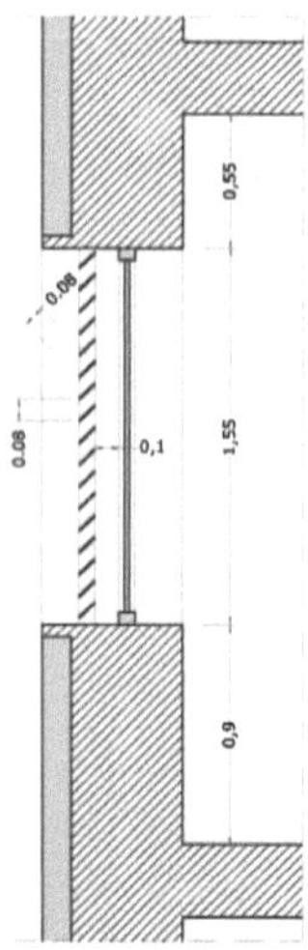
</td>
<td colspan="6">D1.2 - Venetian Blinds (tilting: 0°, 30°, 60°)
Best energy performances for 60° blind tilting: F_s = 47% for south orientation and 44% for west orientation. For every orientation F_W is negligible due to top packaging capability into a very small space.
Adequate light distribution is guaranteed only with south orientation.
This shading does not affect the sound insulation of the façade.</td>
</tr>
<tr><td colspan="6">Synthetic Solar Shading Evaluation</td></tr>
<tr><td>TF</td><td>M</td><td>F_w</td><td>F_s</td><td>VC</td><td>$D_{2m,nT,w}$</td></tr>
<tr><td>☺</td><td>☺</td><td>☺</td><td>☺</td><td>☹</td><td>😐</td></tr>
</table>

Good	**Not relevant**	**Not satisfactory**
☺	😐	☹

Table 10 (continued)

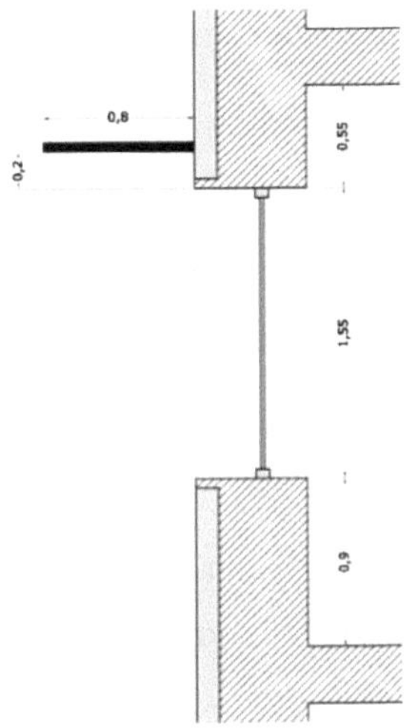

D1.4 - Overhang opaque fixed horizontal
Best energy performances for south orientation: $F_s = 37\%$ and $F_w = 23\%$. For west/east orientation $F_s = 25\%$ and $F_w = 16\%$.
Adequate light distribution is guaranteed for all the orientations. Visual comfort improves especially in summer.
This shading can negatively affect the sound insulation of the façade as it reflects acoustic waves towards the window.

Synthetic Solar Shading Evaluation					
TF	M	F_w	F_s	VC	$D_{2m,nT,w}$

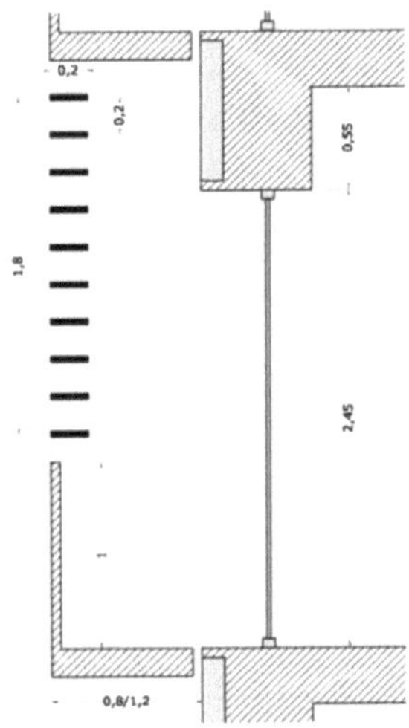

D1.5 - Sunshade vertical with blades tilted integrated with balcony (tilting: 0°, 30°, 60°)
Best energy performances for south orientation: 0° tilting blades imply $F_s = 77\%$ and $F_w = 65\%$. 30° and 60° tilting blades imply a further Qs_s reduction of 20% and Qs_w of 33%.
Adequate illumination is guaranteed only for south orientation and 0° and 30° tilting blades.
This shading with 0° tilting blades, if realized with sound absorbing material, improves the sound insulation of the façade; this performance increases with the highest floors of the building.

Synthetic Solar Shading Evaluation					
TF	M	F_w	F_s	VC	$D_{2m,nT,w}$

Good	Not relevant	Not satisfactory

Table 10 (continued)

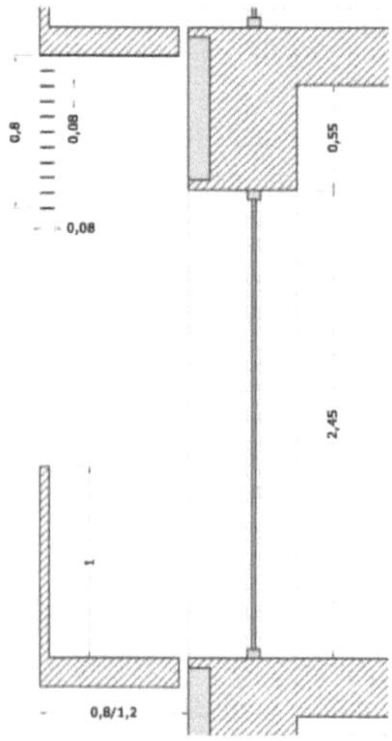

D1.6 - Sunshade vertical with blinds tilted integrated with balcony (tilting: 0°, 30°, 60°)
Best <u>energy performances</u> for south orientation: 0° tilting blades imply $F_s = 64\%$ and $F_W = 43\%$. 30° and 60° tilting blades imply a further Qs_s reduction of 4% and Qs_W of 10%.
Adequate <u>illumination</u> is guaranteed for all the year and different orientations. This shading guarantees a good uniformity of illumination.
This shading with 0° tilting blinds, if realized with sound absorbing material, improves the <u>sound insulation</u> of the façade; this performance increases with the highest floors of the building.

Synthetic Solar Shading Evaluation

TF	M	F_w	F_s	VC	$D_{2m,nT,w}$
☹	🙂	☹	🙂	🙂	🙂

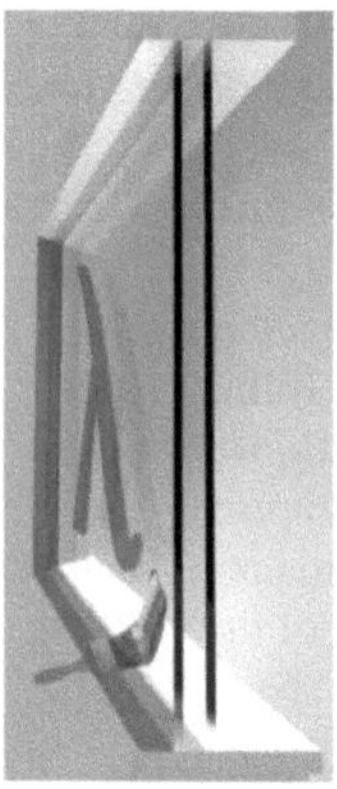

D2.1 and D2.2 - Solar control glasses
<u>Energy performances</u> are similar for winter and summer and different orientation as a function of glass solar factor g. For D2.1 $F_W = 23\%$ and for D2.2 $F_S = 68\%$. Use of solar control glasses can be considered as a proper strategy alternative to external sun shading devices for historical buildings.
Adequate <u>illumination</u> is guaranteed, even if for west/east orientation basic visual tasks are not guarantee for the 30-40% of the year.
Solar glasses can affect the <u>sound insulation</u> of the façade only if Rw of the glass is different from the original.

Synthetic Solar Shading Evaluation

TF	M	F_w		F_s		VC	$D_{2m,nT,w}$
		D2.1	D2.2	D2.1	D2.2		
🙂	😐	😐	☹	😐	🙂	🙂	😐

Good	Not relevant	Not satisfactory
🙂	😐	☹

References

1. Baldini S, Carletti C, Cellai G, Nannipieri E, Sciurpi F, Secchi S (2011) Solar radiation control: acoustic and lighting performances of solar shading devices. In: 48th AICARR international congress Baveno, 22–23 Sept
2. Boyce P (1995) Minimum acceptable transmittance of glazing. Lighting Res Technol 27(3):145–152
3. Boyce P, Hunter C, Howlett O (2003) The benefits of daylight through windows. Lighting Research Center, Rensselaer Polytechnic Institute, Troy, New York
4. Busa L, Cellai G, Secchi S (2005) Acoustic protection of the buildings. Alinea Editore, Florence
5. Busa L, Secchi S, Baldini S (2010) Effect of façade shape for the acoustic protection of buildings. Building Acoustics 17(4):317–338
6. Carletti C, Cellai G, Nannipieri E, Pierangioli L, Sciurpi F, Secchi S (2012) Energy refurbishment of existing buildings with interventions on windows and solar shadings. In: 67th ATI national congress, Trieste, 11–14 Sept
7. Cellai G et al (1994) Energy aspects of windows. In: 49th ATI national congress, Perugia, Italy, pp 271–283
8. Cellai G, Secchi S, Nannipieri E, Baldini S (2011) Acoustic and lighting performances improvement of solar shadings. In: Proceedings of the 38th AIA national congress, Rimini, 8–10 June
9. Cellai G, Secchi S (2009) Evaluating the performance of windows for well-being and reducing energy consumption. In: 47th AICARR international congress system, energy and built environment toward a sustainable comfort, Tivoli (RM) Italy, 8–9 Ottobre, pp 219–230
10. Energy Resources Design (2006) High-performance glazing: making the most of today's advanced technologies. e-News, No. 54
11. Farina A (2000) Validation of the pyramid tracing algorithm for sound propagation outdoors: comparison with experimental measurements and with the ISO/DIS 9613 standards. Adv Eng Softw 31(4):241–250
12. LBNL (US DOE) (2010) EnergyPlus engineering reference, Berkeley, CA, USA
13. Nabil A, Mardaljevic J (2005) Useful daylight factors. Energy Buildings 38(7)
14. Nabil A, Mardaljevic J (2005) Useful daylight illuminance: a new paradigm to access daylight in buildings. Lighting Res Technol 37(1):41–59
15. REHVA (2010) Guidebook n°12. In: Wouter Beck (ed) Solar shading—How to integrate solar shading in sustainable buildings
16. Relux Informatik AG (2007) Relux professional 2007 Manual, Basel, Switzerland. www.relux.ch
17. Russo V (1994) Practical guide-selection and application of special glass insulating. Glass Manufacturers Association

Normative References

18. CIE (1972) Daylighting: international recommendations for the calculation of natural daylight, vol 16. CIE Pub, Paris
19. CIE (1973) Standardization of luminance distribution of clear skies, vol 22. CIE Pub, Paris
20. EN 12207 (1999) Windows and doors—Air permeability—Classification
21. EN 12207 (2000) Windows and doors-permeability-classification
22. EN 12208 (1999) Windows and doors—watertightness—classification
23. EN 12210 (1999) Windows and doors—resistance to wind load—classification
24. EN 12354-3 (2000) Building acoustics—estimation of acoustic performance of buildings from the performance of elements—Part 3: airborne sound insulation against outdoor sound

25. EN 12519 (2004) Windows and pedestrian doors—terminology
26. EN 410 (2000) Glass in building—determination of luminous and solar characteristics of glazing
27. EN ISO 10077-1 (2002) Thermal performance of windows, doors and shutters—calculation of thermal transmittance—simplified method
28. EN ISO 12464-1 (2004) Light and lighting—lighting of work places-indoor work places
29. EN ISO 14683 (2007) Thermal bridges in building construction—linear thermal transmittance—simplified methods and default values
30. UNI 11296 (2009) Acustica—Linee guida per la progettazione, la selezione, l'installazione e il collaudo dei sistemi per la mitigazione ai ricettori del rumore originato da infrastrutture di trasporto (guidelines for the design, selection, installation and testing of systems for the mitigation of noise, originating from transport infrastructure to receptors)
31. Weather data IGDG. http://apps1.eere.energy.gov/buildings/energyplus/cfm/weather_data3.cfm

Improving the Energy Efficiency of Heating Systems in Europe's Residential Buildings

L. Ceccotti, A. De Angelis and O. Saro

Abstract Although the ultimate solution is the complete energy-related renovation of building, upgrades of heating and air-conditioning system components represent the quickest and least costly way to reduce significantly the heating and cooling energy requirements. Such upgrades can be implemented independently of the refurbishment of the building envelope. Numerical simulations of the thermal behaviour of existing buildings have been performed with reference to one-family and multi-family buildings in climate conditions typical of different European countries. The selected building models consider heating systems and envelopes with different thermal characteristics that are customary of existing buildings. A series of tools, such as tables and graphs, are presented in order to enable a rapid and effective preliminary assessment of the benefits rising from the replacement of one or more components of the system. On this basis, it is possible to estimate the increase in the energy efficiency resulting from a given heating system upgrade. For a more complete analysis of the energy-related upgrades, some economic aspects have been considered. In order to assess the cost-effectiveness, typical tools of the financial analysis have been used. Each upgrade has been treated as an investment with an initial outlay and a series of incoming cash flows, which correspond to the annual energy savings from the considered upgrade.

Nomenclature

A	Surface (m^2)
C	Cost (€/kWh)
CF	Cash flow (€)
COP	Coefficient of performance
d	Diameter (m)
ESC	Energy-saving cost

L. Ceccotti · A. De Angelis · O. Saro (✉)
Department of Electrical, Management, and Mechanical Engineering,
University of Udine, Udine, Italy
e-mail: onorio.saro@uniud.it

A. Magrini (ed.), *Building Refurbishment for Energy Performance*,
Green Energy and Technology, DOI: 10.1007/978-3-319-03074-6_3,

f	Factor
F_{i-j}	Shape factor for the surfaces i and j
FC	Load factor
h	Convection heat transfer coefficient (W m^{-2} K^{-1})
HCV	Higher calorific value of the fuel (Wh m^{-3})
HDD	Heating degree hours based on 20 °C (°C d)
i	Interest rate
IRR	Internal rate of return
LCV	Lower calorific value of the fuel (Wh m^{-3})
n	Exponent of the emission system
Q	Quantity of heat (Wh)
r	Annual energy cost escalation rate
R	Thermal resistance (m^2 K W^{-1})
S	Size factor (safety factor)
t	Time (s)
T	Kelvin temperature (K)
U	Thermal transmittance (W m^{-2} K^{-1})
V	Volume of air in a heated zone (m^3)

Greek Symbols

Δ	Prefix for difference
E	Emissivity
Φ	Thermal power (W)
ϑ	Celsius temperature (°C)
σ	Stefan–boltzmann constant

Subscripts

a	Internal air
avg	Average
bw	Back wall
conv	Convection
d	Distribution system
e	External air
E	Energy
el	Electric
em	Emission
ext	External
f	Floor
fw	Front wall
H	Heating
int	Internal
l	Losses
k, j	Indices

max	Maximum
min	Minimum
nd	Needed
nom	Nominal
off	Burner off
on	Burner on
p	Primary
r	Return
rad	Radiation
s	Supply
tot	Total
w	Water
win	Window

1 Introduction

Many European post-war buildings were designed and built in accordance with criteria which generally did not take into account aspects connected with reducing their energy requirements. Most of these buildings have not yet undergone significant renovation work and constitute a building stock characterised by decidedly poor energy performance levels. Hence, it is important to renovate these buildings in terms of energy performance, to reduce the impact of heating-related expenses on the household budgets of their occupants as well as reduce the residential sector's share of the European Union's overall primary energy requirements.

In general, in the process of energy-related renovation of building/heating systems, upgrades to system components constitute the quickest, least costly way to achieve the goal of significantly reducing heating and cooling energy requirements and may be implemented independently of those on the building shell.

This chapter sets out the main upgrades that can be performed on single-family or multi-family residential buildings.

The text that follows has been prepared on the assumption that the reader needs to choose the most cost-effective upgrade for a given building. With this in mind, in order to enable a rapid, effective preliminary assessment of the benefits deriving from the replacement of one or more components of the system, a number of tables have been prepared on the basis of which, after identifying the characteristic configuration of the building under examination, it is possible to estimate the improvement in energy efficiency resulting from a given energy efficiency upgrade.

A complete account of energy-related upgrade measures cannot disregard the economic aspect. Indeed, an upgrade requiring a relatively small outlay that saves a small amount of primary energy may be more cost-effective than one which brings larger energy savings but which also requires a higher initial investment.

In order to assess the cost-effectiveness of upgrades, tools specific to financial analysis have been used, treating each upgrade as an investment characterised by an initial outlay and by a series of incoming cash flows represented by an economic valuation of the annual savings, which the upgrade allows to be achieved.

2 Characteristics of Heating Systems

2.1 Combustion Boilers

Most existing heating systems serving residential buildings consist of gas- or liquid-fuel-based combustion boilers. In order to estimate their efficiency, reference has been made to the classification, reported in Table 1, proposed by European Council Directive 92/42/EC according to their useful efficiency at their rated output, at an average water temperature in the boiler of 70 °C, and operating under partial load conditions of 30 %, with a different temperature according to boiler type.

Standard boilers, which probably constituted the state of the art at the time of construction of the buildings whose heating systems are subject to renovation, can nowadays be defined as low efficiency when compared with low-temperature boilers and even lower efficiency compared with condensing boilers.

For standard boilers, it is advisable not to lower the flue gas temperature below the value which corresponds to their dew point, as this risks damaging the boiler as a result of the formation of condensation, which is particularly corrosive in the presence of sulphuric acid or nitric acid. In particular, Diesel oil, fuel oil or heavy oil boilers must be kept at a temperature whose typical values at the flue outlet are around 120–140 °C because of sulphur content of flue gases. For standard gas boilers, in some cases, the flue gas temperature may fall to lower values.

For the reasons just mentioned, as the flue gas temperature is linked to the average temperature of the water in the boiler, before the advent of "low-temperature" boilers, it was necessary to keep the temperature of the return water from the heating system at a sufficiently high value to prevent the condensation of the water vapour contained in the flue gases. Therefore, in the installation of a standard boiler, it is necessary to create a circuit known as an "anti-condensation circuit". In some cases, this system consists of a three-way valve operated by a servomotor, which is controlled in turn by a sensor which measures the boiler return water temperature or the temperature of the water inside the boiler itself, as shown in Fig. 1. Implementation of solutions of this type requires the boiler temperature to be kept at a high value and close to the rated temperature even when the system is on but does not demand heat energy.

In addition to avoiding the formation of condensation in the parts which make up the boiler, in standard boilers, it is necessary to keep the value of the temperature of flue gases significantly higher than that of the external air temperature when the gases are evacuated using the natural draught method. Indeed, with a

Table 1 Useful efficiency requirements 92/42/EC

Boiler type	Range of power output	Efficiency at rated output		Efficiency at part load	
		Average boiler water temperature	Efficiency requirement	Average boiler water temperature	Efficiency requirement
	(kW)	(°C)	(%)	(°C)	(%)
Standard boiler	4–400	70	$\geq 84 + 2 \log \Phi_{nom}$	≥50	$\geq 80 + 3 \log \Phi_{nom}$
Low-temperature boiler[a]	4–400	70	$\geq 87{,}5 + 1{,}5 \log \Phi_{nom}$	40	$\geq 87{,}5 + 1{,}5 \log \Phi_{nom}$
Gas condensing boiler	4–400	70	$\geq 91 + 1 \log \Phi_{nom}$	30[b]	$\geq 97 + 1 \log \Phi_{nom}$

[a] Including condensing boilers using liquid fuels
[b] Temperature of boiler water supply

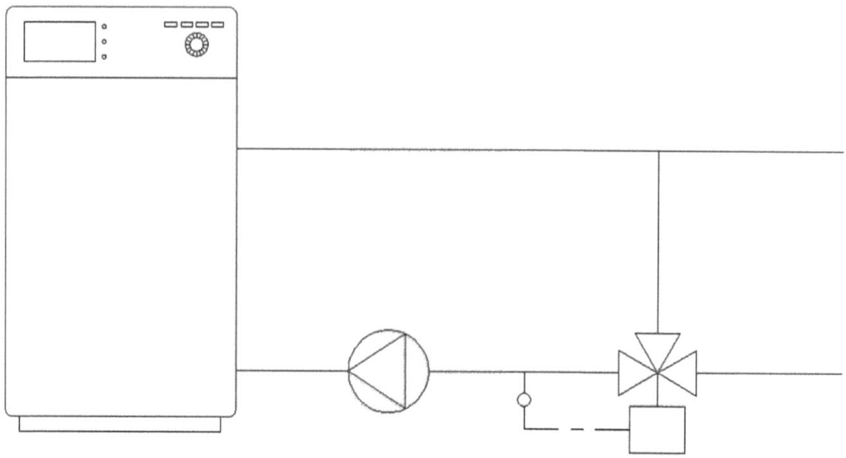

Fig. 1 Simple diagram of an "anti-condensation" circuit for a small boiler

standard, natural gas boiler, in order for the flue gases to have a sufficiently lower density than that of the external air in order to guarantee efficient natural draught in every season, their temperature must be higher than 120 °C.

The characteristics just highlighted mean that the average temperature of the water in the boiler is close to the rated temperature for the entire period of operation with the consequence that they have high energy losses both at the flue and at the boiler casing as a result of convection and radiation.

For a standard boiler installed as part of an old heating system, flue heat losses may range on average between 15 and 20 % of the higher calorific value (HCV). Losses ascribable to the shell can be estimated at around 3 % of HCV. To improve a boiler's performance, it is opportune to reduce the energy contained in the products of combustion. This is why low-temperature boilers and condensing boilers have been introduced on the market. For these kinds of boilers, it is not necessary to create an anti-condensation system, and the boiler can cool freely when the system is demanding no energy or less energy and the outlet temperature can be reduced with a climate-based control system equipped with an external sensor.

A condensing boiler is one in which, under normal operating conditions, and for given water temperatures, the water vapour contained in the combustion products is partially condensed, and it is possible to use for heating purposes a part of its latent heat which, for the temperature levels in question, is approximately 2,500 kJ/kg. In this way, under certain operating conditions, condensing boilers ensure significant energy savings in comparison with standard boilers.

Natural gas (CH_4) is the fuel typically used in these boilers and, of all hydrocarbons, is the one with the highest H/C ratio. This means a larger quantity of water in the flue gases and a greater difference between lower and higher calorific values (approximately 10 %), as it can see in Table 2.

Table 2 Calorific values of a number of gaseous fuels

Fuel	Calorific values at 25 °C				Normal density
	(kJ/kg)		(kJ/m_n^3)		(kg/m_n^3)
	Lower	Higher	Lower	Higher	
Hydrogen	120,000	141,900	10,800	12,770	0.090
Methane	50,050	55,550	35,890	39,830	0.717
Propane	46,350	50,400	93,630	101,800	2.020

Hence, it is more energy efficient to allow the vapour contained in the flue gases—which normally contain water, CO_2, N_2 and negligible traces of other compounds in energy terms—to condense.

The necessary condition for condensation to take place is that the temperature of the return water from the heating system is lower than the dew point of the products of combustion.

The dew point of the vapour contained in the flue gases of a stoichiometric natural gas combustion process is around 59 °C. In practice, combustion always occurs with an "excess of air". The value of the dew point decreases as the excess air increases as a result of the dilution of the flue gases and a lower partial pressure of the water vapour. For a 20 % excess of air, the dew point is just over 55 °C.

Thus, by cooling the flue gases to below this temperature value, condensation is formed.

The colder the return water from the heating system, the lower the temperature of the exhaust gases can be, and the greater the quantity both of condensed vapour and of latent heat recovered. However, condensation of all of the vapour contained in the exhaust gases is never obtained, because as they dry, the vapour pressure and dew point fall.

As a consequence of the cooling of the flue gases, latent heat is recovered and a decrease in both sensible heat losses in flue gas and boiler casing loss.

The temperature of return water from the heating system depends primarily on the type of heating element installed and on the thermal load that needs to be supplied to the building. The latter varies, over the course of the day and during the entire heating period, as a result of changes in external climatic conditions and the contribution of solar and thermal energy gains.

From the point of view of evacuation of combustion products, a condensing boiler expels gases at temperatures gradually decreasing from 80 to 40 °C depending on the season, or the temperature of the return water from the heating system. The density of the exhaust gases is therefore much more variable compared to that of the external air temperature, and in any case, it is lower than that. Such a situation cannot guarantee a sufficient, constant natural draught, especially during the temperate season. For this reason, in condensing boilers, it is necessary to use a fan which ensures that the flue gases have the necessary head of pressure in order to pass both through the boiler and through the evacuation device. Stopping the fan must also mean that the fuel supply is cut.

Nearly, all modern condensing boilers are able to operate by modulating the heating output, with the result that they can adjust to the thermal load demanded by the system [1]. Often, during the heating period, the external temperature is higher than the minimum annual value, the building heat losses are lower than maximum and the boiler is underused in relation to system heating demands.

The modulation range, which is defined as the difference between the maximum heat output and the minimum heat output below which the boiler operates in "on/off" mode, varies according to the boiler's design characteristics.

When the heating system starts, controlled by the room thermostat, the boiler operates at maximum power in order to heat the heat transfer fluid to its maximum operating temperature in the shortest possible time. The boiler continues to operate in this way until the temperature of the water in the system reaches the temperature value required by currently operating conditions. At this point, in order to keep the heat transfer fluid at the temperature required for efficient operation, the boiler is required to supply less heat. Thus, if the boiler is of the modulating type, it can switch to operation with the burner on at a heating output value ranging between the minimum output and the rated output.

In modulating boilers, the flame can be continuously adjusted within a range of power outputs in the modulation range. Hence, a boiler of this type may operate continuously at an intermediate power output between the minimum and maximum, or intermittently at the minimum power output.

The parameter that enables the boiler's operating conditions to be determined is the boiler load factor (FC), which is defined as the ratio of firing time (burner on) to total activation time (which thus includes both firing time and standby time, during which the boiler is in standby mode, but no heat is required by the system).

$$\mathrm{FC} = \frac{t_{\mathrm{on}}}{t_{\mathrm{on}} + t_{\mathrm{off}}} \tag{1}$$

A non-modulating boiler is characterised by a load factor that is always less than one, while for a modulating boiler, the value is one during those periods in which the boiler is in the modulating phase.

It is pointed out that also in the case of a modulating condensing boiler, in the absence of a weather-based climate control system equipped with an outdoor sensor, the radiators operate continuously at the design temperature, regardless of the thermal load required by the space to be heated. Lower heat requirements, such as during the early and latter months of the heating season, do not result in lower heat transfer fluid temperatures, but rather in more rapid boiler operation times both at maximum output and at minimum output. Thus, the boiler often operates under modulation and standby conditions [1].

Fig. 2 Functional diagram of a refrigeration cycle

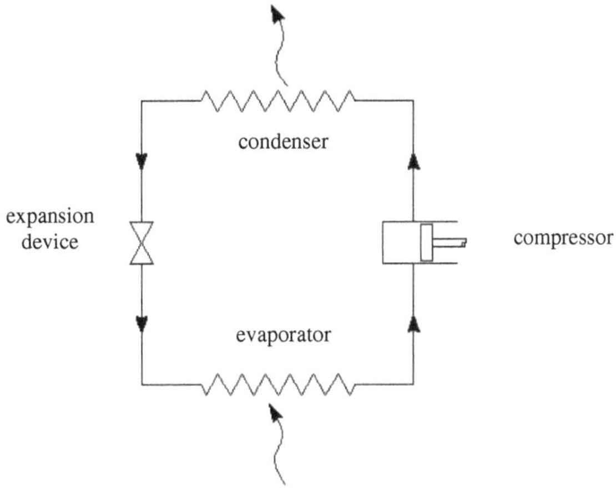

2.2 *Electrical Heat Pumps*

For some years now, there has been growing use of electrical heat pumps for winter climate control purposes in residential buildings.

An electrical heat pump's operation is based on the reverse thermodynamic cycle, which enables energy to be transferred from a lower-temperature source to one at a higher temperature through the supply of work. The practical effect is the supply of heat to a system, which is at a higher temperature than the ambient temperature.

By simplifying the diagram of the thermodynamic cycle of a heat pump down to its essential components only, it can be represented as in Fig. 2, which shows

- the compressor: the component in which the refrigerant fluid is compressed, at the cost of useful energy as work supplied from the outside;
- the condenser: a heat exchanger which transfers heat energy to the hot source, which may consist, for example, of the air in the space to be heated, or of the heat transfer fluid in the case of water-based systems. As a result of this heat exchange, which therefore corresponds to the useful heat flow, in the condenser, the refrigerant is cooled to the condensation temperature, and the consequent gas-to-liquid phase transition is obtained;
- the expansion device: consisting generally of a valve in which a non-reversible adiabatic expansion of the refrigerant fluid takes place, until it reaches a pressure value which, at least theoretically, is the same as the pressure value at the compressor inlet;
- the evaporator: a heat exchanger in contact with the cold source, which may be constituted by the air in the external environment, or by a fluid which in turn exchanges heat with a water source or with the ground. Inside the evaporator, the refrigerant fluid undergoes a liquid-to-gas phase transition and in this transformation absorbs heat from the cold source.

In order for the heat exchange between the refrigerant fluid and the hot source to take place in the condenser, the condensation temperature needs to be higher than the temperature at which the system is kept heated. Similarly, in order to take place, heat exchange in the evaporator requires the temperature of the refrigerant fluid at this point in the cycle to be lower than the ambient temperature (either of the water or of the ground) with which the exchange takes place. The difference between the condensation and evaporation temperatures influences the efficiency of the thermodynamic cycle. In fact, considering an ideal Carnot cycle operating between the condensation temperature, T_c, and the evaporation temperature, T_e, the coefficient of performance (COP) is thus given by

$$COP_{Carnot} = \frac{T_c}{T_c - T_e} \tag{2}$$

The COP_{Carnot} coefficient represents the maximum efficiency theoretically obtainable with reversed thermodynamic cycle operating as a heat pump. In order to maximise the COP_{Carnot} coefficient, it is necessary to reduce the (T_c−T_e) difference to a minimum. This can be achieved by keeping the temperature of the hot source as low as possible and therefore if the latter is the same as that of the heat transfer fluid in a heating system, by using systems equipped with heating elements sized to operate at a low temperature.

In order to increase the efficiency, it is also possible to act upon the cold source by selecting it so that it is available at the highest possible temperature and also constant during the heating season. It is for this reason that heat exchange systems with alternative sources to external air have become widespread, such as ground or lake water.

If the machine's condenser transfers heat to a heat transfer fluid consisting of water, as in the case of heating systems based on radiant heating panels, radiators or fan coil units, the heat pumps will be, respectively, of the air-to-water, water-to-water or ground-to-water type. In terms of operating principles, these machines are very similar to each other. However, there are substantial differences in their design characteristics, in the works required for their installation and in how they are integrated into the heating system, in their efficiency and in their cost.

The heating output of a water-to-air heat pump used for heating purposes is affected by the temperature of the cold source: as the external temperature decreases, so does the heating output which the machine is capable of providing. Conversely, the heat energy requirements of a building increase as the external temperature decreases. For some models of heat pump, it may also be the case that the maximum output temperature of the heat transfer fluid is not guaranteed for the lowest external temperature values. This may lead to problematic situations, especially in cases where the heat pump is used both for heating and for supplying hot water for domestic use.

A system designer, who plans to provide a building with its entire heat energy requirements using a heat pump—that is, without recourse to an auxiliary heater—will be impelled to choose such a machine size, as it will guarantee the supply of the rated design output even when the external temperature falls to the lowest levels. In this case, the machine may be significantly oversized in respect of average required load conditions and hence the need to regulate the amount of energy supplied. This can be achieved by activating and deactivating the compressor (using an "on/off-type" regulation) or by modifying the air volumetric flow rate of the compressor itself. The second strategy can be implemented by using refrigeration units (operating as a heat pump) with one or more zoned circuits, or by equipping the heat pump with an inverter which makes it possible to adjust the flow rate of refrigerant fluid by adjusting the compressor speed, external climatic conditions being equal, resulting in an increase in the machine's COP. Also, with a refrigerant fluid flow rate regulation system, a minimum value exists below which the output can no longer be regulated, and the compressor operates in on/off mode.

From the design type perspective, air-to-water heat pumps are commercially available in a monoblock configuration, used primarily in medium-to-large-sized systems, and with the option of connecting more than one boiler on the same system, in order to split the installed heating output. Other models consist of an external unit and an internal unit, which are used, particularly, in cases where lower heating output is required.

In addition, it is necessary to consider the characteristics of the heating system, which will be receiving the heat transfer fluid from the machine that is to be installed. In particular, both efficiency and capacity vary according to the temperature to which the fluid is heated.

2.3 Heat Emission Systems

By far, the most widely used heat emission system used in existing buildings is hot water radiators. A small percentage of buildings are heated by systems based on radiant heating panels.

A very small number—compared with radiator-based systems—of single-family buildings feature convector heaters and fan coil units. The latter are mostly present in single-household buildings with heating systems of the mixed type, in which the heat emission device with heat exchange characterised by a main convective component is used to heat rooms used occasionally and therefore requires the air temperature to be raised very rapidly.

2.3.1 Radiators

Contrary to what their name might lead one to expect, radiators supply the space in which they are located with heat energy mainly by means of convection, while they exchange approximately 30 % of their total output by means of radiation. This percentage may vary according to the type of radiator and is higher in those of the “plate” type compared with so-called column radiators. Only the front surface provides radiation, while the convective component must be considered to be the entire surface of the heat-emitting body.

The sizing of the heat-emitting body to be installed in a room is carried out according to the thermal output, i.e. the rated power usually indicated in the manufacturer’s catalogue. This value is obtained from tests conducted in a laboratory following standardised procedures and is expressed as a function of the difference between the average water temperature and the ambient temperature. Assuming that the radiators are supplied with water at a temperature of $\vartheta_s = 75$ °C, which on returning to the boiler is at a temperature of $\vartheta_r = 65$ °C, and that the ambient temperature is $\vartheta_a = 20$ °C, the temperature difference between the average water temperature and the ambient temperature $\Delta\vartheta_{nom}$ is equal to

$$\Delta\vartheta_{\mathrm{nom}} = \frac{(\vartheta_s + \vartheta_r)_{\mathrm{nom}}}{2} - \vartheta_a = 50\,\mathrm{K} \tag{3}$$

In order to determine the thermal output of radiators under different operating conditions from test conditions, the following relation is used:

$$\Phi = \Phi_{\mathrm{nom}} \left(\frac{\Delta\vartheta}{\Delta\vartheta_{\mathrm{nom}}} \right)^n \tag{4}$$

where n is an exponent which is determined experimentally in the laboratory and is stated in the catalogues, while the temperature difference is expressed as follows:

$$\Delta\vartheta = \frac{(\vartheta_s + \vartheta_r)}{2} - \vartheta_a \tag{5}$$

One of the causes of inefficiency in heat emission devices is the presence of convection flow, which is more or less significant depending on the mode of heat exchange that characterises each heat-emitting body and which results in a non-uniform distribution of the ambient air temperature. Convection flow, and heat exchange via radiation with the coldest surfaces of the building envelope, leads to an increase in the internal surface temperature of the walls, especially in radiator-based heating systems, with a resulting increase in the interior surface heat transfer coefficient and therefore transmittance. Thermal resistances involved in heat exchanges between thermal emission devices and building envelope are show in Fig. 3. In evaluating the heat-emitting performance of a radiator, it is necessary to estimate the surface area, A, of the heat-emitting bodies with reference to the building’s maximum dispersion value and assuming that heat energy is supplied by the radiators partly through convection Φ_{conv} and party through radiation Φ_{rad}. Estimating that Φ_{conv} accounts for approximately 70 % of the rated output of the heat-emitting body, it is possible to estimate an area, A_{conv}, associated with heat exchange via convection and an area, A_{rad}, associated with heat exchange via radiation, by the following equations:

$$A_{\mathrm{conv}} = \frac{\Phi_{\mathrm{conv}}}{h(\vartheta_{s,\max} - \vartheta_a)} \tag{6}$$

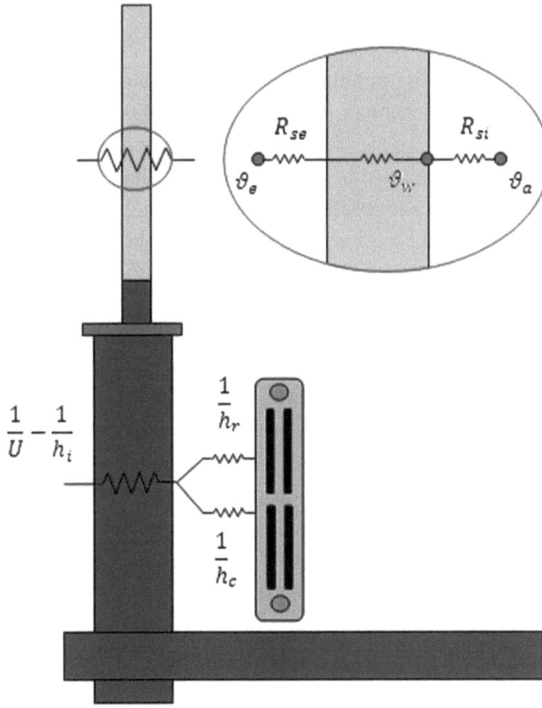

Fig. 3 Representation of the resistances involved in heat exchange with the glass surface and with the wall behind the radiator

$$A_{rad} = \frac{\Phi_{rad}}{\varepsilon\sigma\left(T^4_{s,max} - T^4_a\right)} \tag{7}$$

A_{rad} is considered to be split between a portion representing heat exchange via radiation with the wall on which the radiator is positioned, $A_{rad,bw}$, and a portion representing heat exchange with the rest of the walls in the room, $A_{rad,fw}$. It is assumed that the front and side area of the radiator is responsible for radiation heat exchange with the room around 60 % of the total radiating area of the heating element. The radiator's heating output is calculated for each hour of system operation, and the emission performance is taken to be the ratio of the heat output actually supplied to the space being heated to the required heat.

The energy emitted by the heating device differs from the energy required as a result of losses $Q_{l,em}$, the sum of losses due to convection $Q_{l,em,conv}$ and due to radiation $Q_{l,em,rad}$:

$$Q_{l,em} = A_{rad,bw}(\vartheta_e - \vartheta_a)\left(\frac{1}{U} - \frac{1}{h_{int}} + \frac{1}{h_{conv} + h_{rad}}\right) \tag{8}$$

2.3.2 Radiant Heating

Radiant heating panels are heat emission devices, which, in the room in which they are installed, exchange heat primarily by means of radiation.

With regard to heating applications in residential buildings, reference is implicitly made to low-temperature radiant panels, which work with a maximum outlet water temperature of approximately 45 °C.

These heat emission devices may be located in the floor, the walls or the ceiling. When installed under the floor, they can be buried in a concrete screed and thermally isolated from the structures to make downward heat dispersion negligible.

Heat energy is supplied to the rooms from the hot surface of the floor or the walls, mainly by means of radiation and secondarily by means of convection. Heat emission losses can be estimated, to a good approximation, by evaluating the heat energy exchanged via radiation between the panels and the glass surfaces, $Q_{\text{f-win}}$, since the surface temperature of the latter is significantly lower than that of the opaque walls involved in the radiative heat exchange with the panels. The heat energy exchanged via radiation between the panels and the glass surfaces can be calculated using the following relation:

$$Q_{\text{f}-\text{win}} = \frac{A_{\text{win}}\sigma\left(T_{\text{f}}^{4} - T_{\text{win}}^{4}\right)}{\frac{1-\varepsilon_{\text{f}}}{A_{\text{f}}\varepsilon_{\text{f}}} + \frac{1}{A_{\text{f}}F_{\text{f}-\text{win}}} + \frac{1-\varepsilon_{\text{win}}}{A_{\text{win}}\varepsilon_{\text{win}}}} \tag{9}$$

where the emissivity of the floor surface, ε_{f}, is assumed to be 0.90, while the emissivity of the glass surface, ε_{win}, is assumed to be 0.85.

The surface temperature of the windows was calculated using the following relation:

$$\frac{\vartheta_{\text{a}} - \vartheta_{\text{e}}}{R_{\text{tot}}} = \frac{\vartheta_{\text{a}} - \vartheta_{\text{win}}}{\frac{1}{h_{\text{int}}}} \tag{10}$$

2.4 Regulation Systems

The first type of automatic regulation system historically used in heating systems regulated boiler operation by means of a room thermostat, which controlled ignition of the burner. A control system of this kind, especially in cases in which the boiler is of the standard type, in particular floor-standing boilers, entails large variations in ambient temperature (the variable regulated). This is due to the fact that the boiler–system combination is characterised by the high degree of thermal inertia of the regulation system (comprising the boiler, transmission system and heating elements) compared to that of the system being regulated. This type of regulation is also called on/off.

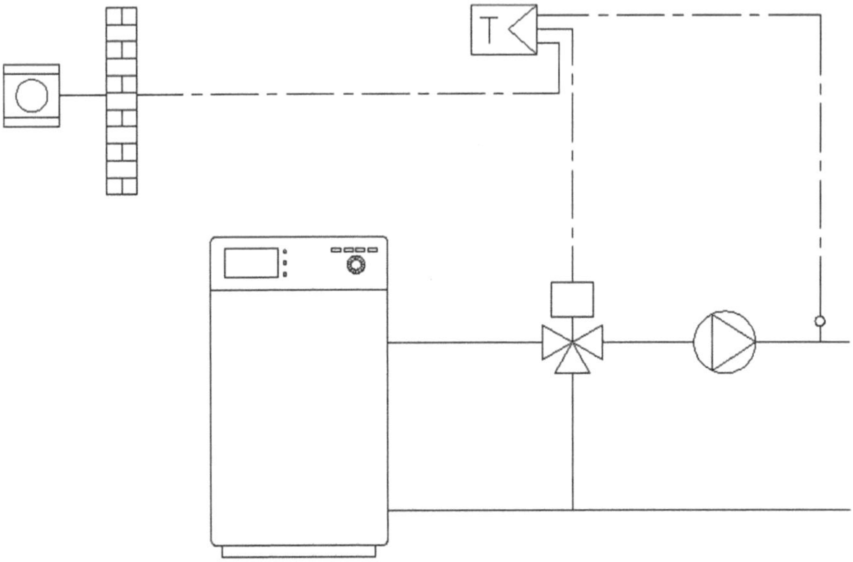

Fig. 4 Diagram illustrating weather-based regulation

In the case of centralised heating systems, regulation systems exclusively permit the temperature of the inlet heat transfer fluid to be controlled. The disadvantages of this are that it is not possible to balance changes in heating load caused by internal heat gains and solar radiation and that such systems are not able to regulate heat emission independently according to the different demands of individual dwellings. Usually, controls of this type are implemented by means of a weather compensator, which receives a signal from a temperature sensor installed on an outside wall of the building and operates by varying the opening position of a three-way mixing valve operated by an actuator installed on the valve. Boiler manufacturers' instructions agree in recommending that this sensor should not be installed in such a way as to be exposed to direct solar radiation, preferably on a north-facing wall. With regulation of this type, the operating point of the system is identified on a curve in which the inlet water temperature is a function of the external temperature and is parameterised according to the characteristics of the type of heating element as provided by its manufacturer.

The temperature sensor, positioned on the inlet pipe after the circulating pump, enables the effect of the action on the three-way valve to be evaluated. A diagram of this type of circuit is provided in Fig. 4.

Weather-based control systems bring benefits both in terms of energy and in terms of thermal comfort, if they are combined with a system for regulating air temperature in individual zones, or better still in individual rooms. In addition, weather-based regulation applied to condensing boilers makes it possible to increase the number of hours during which condensation takes place, as at intermediate outside temperature, a lower outlet temperature is achieved compared with rated conditions, with a consequent increase in boiler efficiency.

With a radiator-based heating system, sized in the traditional manner, it will be the case that in the least favourable weather conditions, corresponding to the minimum outside air temperature during the heating period, water will be sent to the heating elements at a temperature of approximately 75 °C. As a result, the outlet water temperature will assume, under the same conditions, a value close to 65 °C. In fact, the nominal heat output of the radiators is obtained because of a difference between inlet and outlet temperatures of 10 °C and a difference between the average temperature of the heating element and the air temperature of 50 °C. As the outside temperature increases, a linear reduction is assumed in the temperature of the water entering the radiators, down to a minimum value of around 60 °C, which occurs when the outside temperature reaches its highest value during the heating period. If the boiler is of the condensing type, the inlet and outlet temperatures may vary freely according to the heating load required.

As the outside temperature changes, the values of the temperatures of the water sent to the heating elements can be determined using the following relation:

$$\frac{\vartheta_{\mathrm{e}} - \vartheta_{\mathrm{e,min}}}{\vartheta_{\mathrm{e,max}} - \vartheta_{\mathrm{e,min}}} = \frac{75 - \vartheta_{\mathrm{w,s}}}{75 - 25} \tag{11}$$

while for the return temperature to the boiler, the following relation may be used:

$$\frac{\vartheta_{\mathrm{e}} - \vartheta_{\mathrm{e,min}}}{\vartheta_{\mathrm{e,max}} - \vartheta_{\mathrm{e,min}}} = \frac{65 - \vartheta_{\mathrm{w,r}}}{65 - 25} \tag{12}$$

For an underfloor radiant system, the outlet and inlet temperatures are constrained by the maximum limit set on the floor temperature. Therefore, in the following calculations, it is assumed that the underfloor system has been sized in such a way that in the least favourable weather conditions, determined by the minimum outside air temperature, the return temperature to the boiler does not exceed 40 °C. The outlet temperature under the same weather conditions, given a maximum temperature change in the water of 5 °C, will be 45 °C.

As the outside temperature increases, it is assumed that the system of regulation operates by means of a linear reduction in water temperatures until they reach a minimum value, which is slightly higher than the inside temperature of the room being heated when the heating load is very low.

In this case, the pattern of change in the outlet and inlet temperatures, which are obtained according to the outside temperature, may be determined using the following relations:

$$\frac{\vartheta_{\mathrm{e}} - \vartheta_{\mathrm{e,min}}}{\vartheta_{\mathrm{e,max}} - \vartheta_{\mathrm{e,min}}} = \frac{45 - \vartheta_{\mathrm{w,s}}}{45 - 20} \tag{13}$$

for the outlet temperature and

$$\frac{\vartheta_{e} - \vartheta_{e,min}}{\vartheta_{e,max} - \vartheta_{e,min}} = \frac{45 - \vartheta_{w,r}}{45 - 20} \quad (14)$$

for the inlet temperature.

With regard to regulation of the air temperature in the rooms being heated, in buildings with independent heating systems, particularly in single-household buildings, zoned (night and day) temperature regulation systems are common.

While simple regulation using a zone thermostat already permits energy savings thanks to an initial level of automatic heat flow regulation, it absolutely cannot satisfy with precision the requirements of the rooms, which are often subject to different conditions. A thermostat installed in one room cannot adequately take into account the free heat gains—whether internal or external in origin—that exist in the rooms, adjacent to the room whose temperature is being regulated. This means that during some hours of the heating period, an excess amount of heat energy is supplied to some rooms and is thus largely wasted. This aspect leads to a lower value in terms of regulation efficiency for those systems, which are unable to respond accurately and rapidly to changes in heat energy demands.

Another aspect to consider is the possible need for different temperatures for certain rooms. In such cases, it can be assumed that the user can act manually upon the radiator's shutoff valve. However, it would not be possible to achieve the efficiency of automatic regulation in this way.

Regulation systems, which allow the temperature to be controlled automatically and independently in each room, are still uncommon. Such a system can be built in cases when the heating elements are radiators, by installing a thermostatic sensor on the valve controlling the flow of hot water to each heating element. These sensors, which are also known as thermostatic valves, make it possible to keep the ambient temperature at the chosen level, exploiting any free heat gains. The user can set the desired temperature by turning the knob, which displays a graduated scale. Regulation takes place automatically by adjusting the flow of hot water through the radiator by modulating the valve opening according to the air temperature measured by a liquid sensor.

As the temperature measured by the sensor increases, the fluid which it contains expands, pushing on the mechanism coupling it to the valve shutter and moving it towards the closed position. The opening or closure of the shutter varies in proportion to the temperature of the room.

In order for the thermostatic valve to be able to correctly fulfil the function assigned to it, the ambient temperature measured by the sensing bulb must be representative of the temperature for the whole room. It is advisable to prevent it from being heated via convection by the pipes bringing water to the radiator themselves. The air in the room must come into contact freely with the sensing bulb, avoiding exposure to the sun's rays. In some situations, to prevent the erroneous temperature being recorded by the system, thereby rendering its installation ineffective, it is possible to use thermostatic valves with sensing bulbs that can be installed remotely by means of a capillary tube, which enables communication between the bulb and the valve body.

If a zone thermostat and thermostatic valves are installed on the same system, it is advisable to avoid installing the thermostatic sensors in the same rooms of the thermostat, which regulates the zone. In addition, appropriate provisions will need to be adopted to take into account the effect deriving from the inclusion of thermostatic valves on the flow rates of water circulating in the system.

For buildings heated using low-temperature radiant systems, individual room temperature regulation can be achieved by means of thermostats installed in the various rooms and connected to electrothermal actuators. These devices, applied to the shutoff valves with which the distribution collector is equipped, control the circulation of the heat transfer fluid in the circuits of the panels.

2.5 Distribution Systems

The most common distribution systems are the forced-circulation type. Therefore, this is the type considered "as-built" in the simulations conducted.

An ideal distribution system, which enables the heat transfer fluid to be maintained at a constant temperature by the boiler at the heating element, would require a smaller quantity of energy than is necessary in a real-world system. The ratio of these two quantities is a measure of the efficiency of the distribution system.

In order to estimate energy losses in the distribution sub-system, the following relation is used:

$$Q_{\mathrm{l,d}} = \sum_{\mathrm{j}} L_{\mathrm{j}} \cdot U_{\mathrm{j}} \cdot \left(\vartheta_{\mathrm{w,avg,j}} - \vartheta_{\mathrm{a,j}}\right) \cdot t_{\mathrm{j}} \tag{15}$$

As it is not possible to establish one specific type of extension of the distribution system, the length of the system is assumed to be a variable parameter as a function of the heating load required, thus describing a situation in which the heat transfer fluid distribution circuits are activated as the heating power required by the building increases.

The linear thermal transmittance, U_{j}, of the jth distribution circuit can be determined using the following relation:

$$U_{\mathrm{j}} = 0{,}143 + 0{,}0018 \cdot d_{\mathrm{ext}} \tag{16}$$

The temperature of the water in the circuits is assumed to be the average of the outlet and inlet temperatures, while the average temperature of the environment in which the circuits are installed is assumed to be equal to the design ambient temperature.

The activation time of the circuits is assumed to be equal to the activation time of the heat generator.

Table 3 Transmittance values used in the simulations

	U Type A (W/m^2K)	U Type B (W/m^2K)
External wall	0.4	1.2
External ceilings	0.4	1.2
External floors	0.6	1.2
Doors and windows	2.5	4.5

3 Simulations

In order to estimate the potential energy savings that can be achieved through upgrades involving the replacement or installation ex novo of some of the components of the heating plant sub-systems in existing buildings located in several European cities, a number of simulations were conducted, using analytical methods.

Each simulation involves the calculation of the building's primary energy requirements both in an as-built state and in a given as-planned state. The as-built state corresponds, for a building, to a certain combination of heating system components prior to energy-related renovation. An as-planned state consists of a combination of system components following energy-related renovation which entails modifications to one or more of the sub-systems which make up the system.

Using EnergyPlus, a software application for dynamically simulating the energy behaviour of buildings, the hourly heat energy requirements were calculated. It was necessary to define the geometrical and thermophysical properties of the building envelope, in addition to the climatic conditions.

The geometrical properties of each building are described in brief in the corresponding section [2].

The transmittance values of the building envelope components are provided in Table 3.

In order to assess the impact of local climatic conditions on the effects of each energy efficiency upgrade, the simulations were conducted using the characteristic climate parameters for a number of northern, central and southern European cities as inputs. Specifically, the locations taken as reference were Oslo, Dusseldorf, Madrid and Lisbon.

After obtaining the value for the hourly heat energy requirements, with the support of additional analytical procedures, the behaviour and thus the efficiency of heating systems were assessed.

The simulations were conducted for all of the building/heating systems on their as-built state and subsequently in their as-planned state, for comparing the impacts of the upgrades in terms of energy savings. The improvement in energy efficiency that can be obtained was expressed in percentage terms as the ratio of the difference in the primary energy requirements in the two states to that in the primary energy requirements in the as-built state.

The results obtained from the simulations were grouped according to building type and organised in such a way as to group together the ones relating to upgrades involving a modification primarily to the regulation sub-system and those relating to the heat generator.

In the graphs of the results, the symbol $\Delta Q_{p,H}$ is used to indicate the primary energy saving achievable with each upgrade.

3.1 Definition of As-Built States

Both for multi-family and single-family residential buildings, it was assumed that thermal energy for heating is produced using a natural-gas-fired combustion boiler.

The boilers are of the standard type in the case of multi-family buildings, while for single-family buildings, in addition to the standard boilers, low-temperature boilers were considered.

The heating elements are radiators or underfloor radiant panels.

The regulation systems allow the outlet water temperature to be controlled by means of a simple boiler thermostat or weather-based regulation, in the case of buildings with independent heating systems. In centralised heating systems, it was assumed that weather-based systems are installed. The average boiler temperature was considered to be constant and equal to the design temperature, except in the case of low-temperature boilers, for which operation at average modulating temperature was assumed.

For the regulation of the ambient temperature, systems with a zone thermostat were considered for both single- and multi-family residential buildings. For the latter building type, the case of exclusively weather-based regulation, and hence without any control over the ambient temperature, was taken into consideration.

3.2 Definition of As-Planned States

In defining upgrades involving boiler replacement, it was assumed that, following the energy efficiency upgrade performed on the heating system, the heat energy is produced with a gas-fired condensing boiler or with an electrical heat pump of the air-to-water type.

Modifications to the heat emission sub-system between the as-built and as-planned states were not considered. The choice stems from the consideration that any replacement of the heating elements would often make substantial building work necessary. Therefore, in these cases, it would not be correct to assume work on the heating system and not other construction work, which would lie beyond the scope of the objectives initially set.

This type of choice is compatible with the considered upgrades. Indeed, particularly with regard to condensing boilers, the optimal combination from the energy-saving perspective is a radiant panel system operating with water at a low temperature. Nevertheless, real-world experience and numerical evaluations available in the technical literature have shown that excellent results can also be achieved when the heating elements consist of traditional radiators, if care is taken to install a suitable system for regulating the temperature of the heat transfer fluid [3].

In such an operating mode, for heating loads that are lower than system capacity, the conditions for condensing the water vapour contained in the flue gases can arise and thus permit a significant energy saving.

Also in combination with the electrical heat pumps, the type of heating element which guarantees the best results in terms of energy is represented by low-temperature radiant panel systems. Heat pumps which allow heating up to a temperature of 80 °C have been commercially available for several years, albeit at the cost of a significant reduction in the COP (between 20 and 25 % lower in the case of an air-to-water heat pump) compared with the supply of equal heat output at a temperature of 45 °C.

Therefore, in the case of existing radiator-based systems, it is appropriate to verify the sizing of the heating elements installed and, if possible, evaluate whether they should be replaced with others with a higher nominal output, which can operate at a lower outlet temperature compared to the ones already installed. In the case of single-family buildings, the possible replacement of the gas-fired boiler with an electrical heat pump in both radiator-based and radiant panel systems was considered. In the case of multi-family buildings, the heat pump was considered in the as-planned states only in combination with a low-temperature radiant panel system.

Given that the heat distribution system is often enclosed in the walls or in the floors, and that work to renew the heat transfer fluid feed pipes and their insulation would in many cases be difficult to implement without concomitant building work, it was assumed that between the as-built and as-planned states, the distribution sub-system would not be modified.

Any reductions in distribution-related energy losses may derive from changes in the temperature of the heating water, which in turn are brought about by the replacement of the boiler with one which allows a variable average temperature of the heat transfer fluid.

With regard to regulation systems, those systems that enable the greatest efficiency, that is to say individual room temperature control, were considered, also in combination with weather-based regulation systems for varying the boiler outlet water temperature according to the outside temperature.

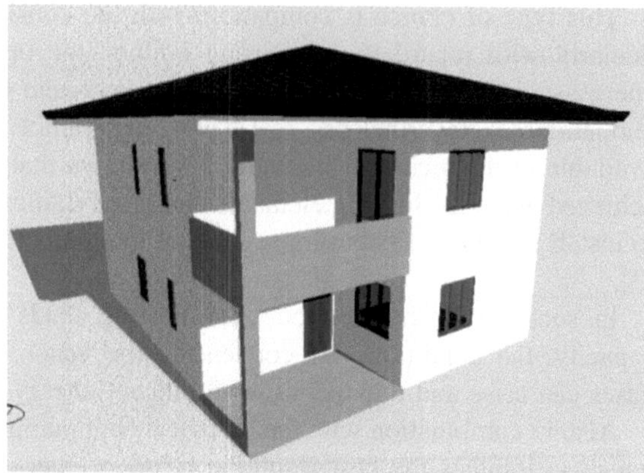

Fig. 5 Two-floor single-family building

4 Results

4.1 Single-Family Building

The results obtained for a single-family building are presented below. The building model is represented in Fig. 5.

The building, with two above-ground floors, has a floor area of approximately 140 m^2, a volume of approximately 430 m^3 and a heat-dispersing surface to heated gross volume ratio (S/V) equal to 0.78 m^{-1}.

The effects of the upgrades to the heating system were evaluated both in the case of a building envelope with thermal transmittance values corresponding to type A and in the case of a building envelope with thermal transmittance values corresponding to type B in Table 3. In the legends of the graphs, these results are indicated using the label U type A and U type B.

The simulations were conducted both for the case of a radiator-based heating system and for the case of low-temperature radiant panels, supplied by one or several distribution manifolds, equipped with shutoff valves for each circuit.

The results were represented in graphs differentiated according to type of heating element, so that they can be more easily consulted and if necessary compared.

4.1.1 Upgrades to the Regulation System

The analysis was carried out to assess the impact of primary energy requirements of a modification on the regulation system, which in the as-built state is assumed to be a zone thermostat for controlling the ambient temperature.

The as-planned states assumed, in the first case, the installation of a system to control the inlet water temperature of the heating elements of a weather-based

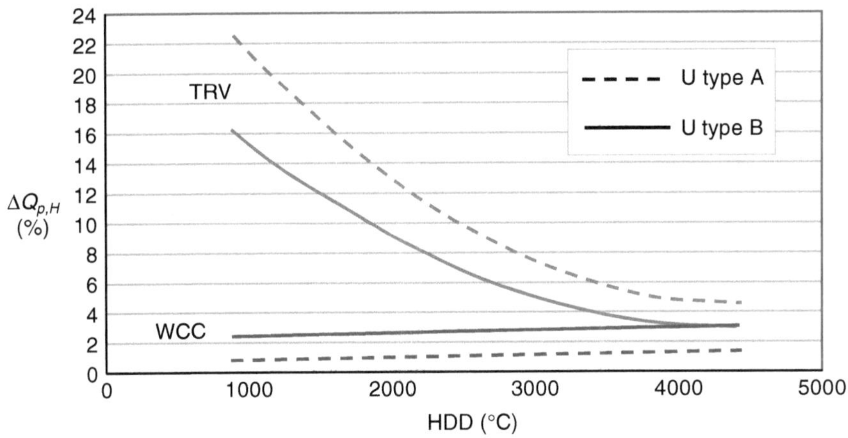

Fig. 6 Radiator system. Installation of WCC or TRV. Standard boiler

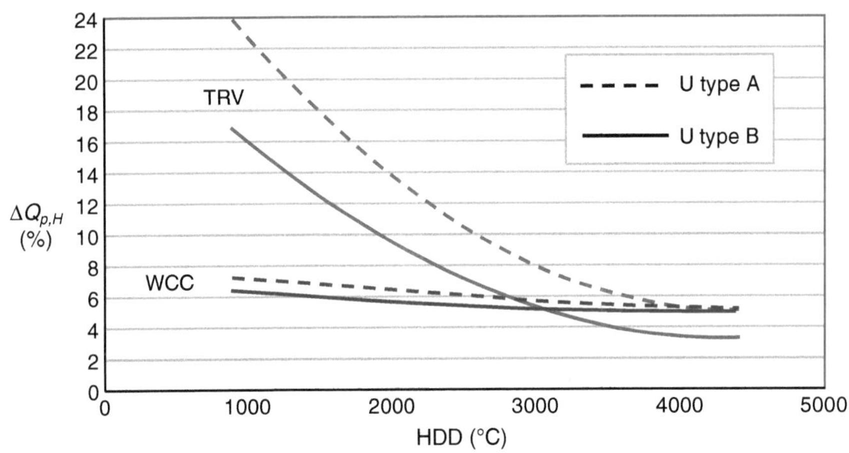

Fig. 7 Radiator system. Installation of WCC or TRV. Low-temperature boiler

control system with an external sensor and in the second case, the installation of an air temperature control in each room.

Figure 6 refers to a standard combustion boiler, Fig. 7 represents a low-temperature boiler, while the results displayed in Fig. 8 were obtained by considering a condensing boiler.

The advantages deriving from the use of a weather-based climate control system are greater in the case of a boiler of the low-temperature type or condensing type than in the case of an upgrade to a heating system fuelled by a standard boiler. This can be attributed to the fact that in the latter case, it is not possible to assume operation at average modulating temperature, and the change in the water temperature is obtained via a mixing process carried out externally to the boiler.

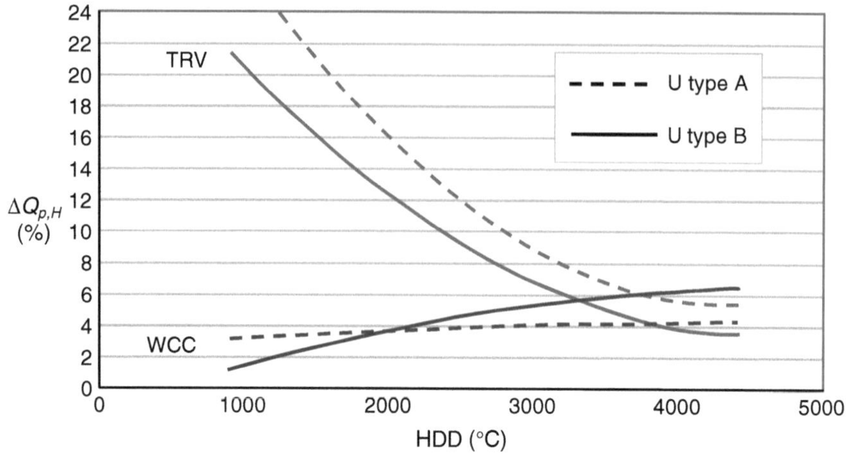

Fig. 8 Radiator system. Installation of WCC or TRV. Condensing boiler

Therefore, boiler energy losses do not decrease when the water temperature in the heating elements is reduced. The positive effects deriving from the decrease in the outlet temperature occur in this case with a reduction in energy losses in the distribution system. In addition, it was observed that the saving is larger when the boiler is of the low-temperature type. This can be attributed to the high-efficiency operation of a condensing boiler in the as-built state as well as in the as-planned state.

The advantage deriving from individual room temperature control using thermostatic sensors can be attributed to a more efficient use of free heat gains, particularly solar radiation via the glass surfaces in the room in which the heating element is installed, compared with what can be achieved using zoned control systems.

The energy saving obtained by using room-by-room regulation increases with warmer climatic conditions and higher solar radiation.

The simulations regarding upgrades to the regulation system, in the case of a low-temperature radiant panel heating system, produced the results shown in Figs. 9, 10 and 11, which are very similar to those obtained for radiator-based systems.

Individual room temperature regulation is achieved by installing a thermostat in each room and connecting these devices to electrothermal actuators which control the closure of the valves on the distribution manifold to which the circuits of the radiant panels are connected.

4.1.2 Upgrades to the Generation System

The results obtained from the simulations of upgrades involving the replacement of the boiler are illustrated below. For each of the cases considered, several

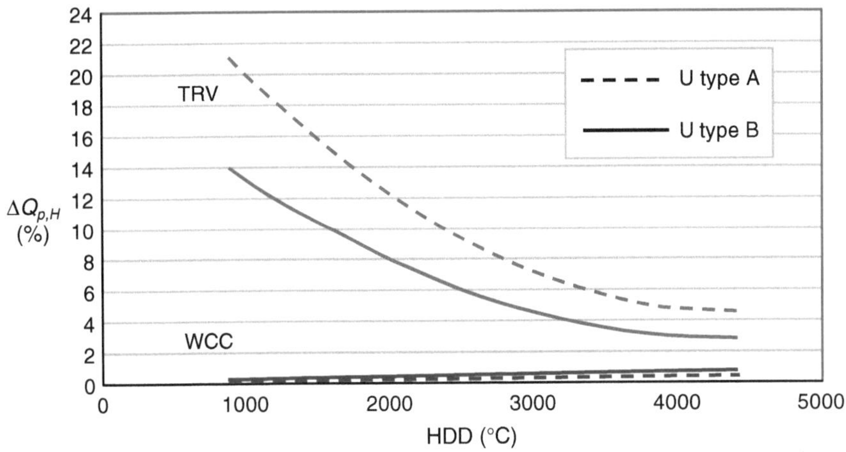

Fig. 9 Radiant panel system. Installation of WCC or TRV. Standard boiler

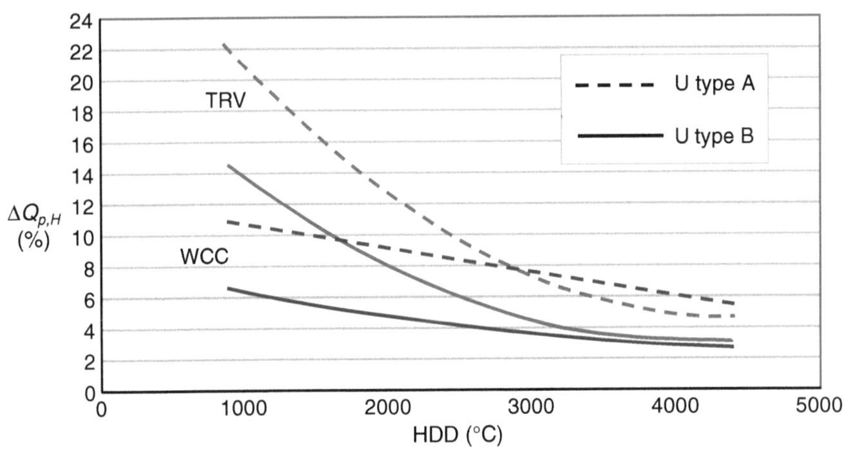

Fig. 10 Radiant panel system. Installation of WCC or TRV. Low-temperature boiler

simulations were conducted: in the first (I in the graph), it was assumed that no modifications are made to the regulation system; in the second (II in the graph), it was assumed that a weather-based water control system is installed; in the third (III in the graph), it was assumed that both a weather-based outlet water temperature control system and individual room temperature regulation were added; the fourth case (IV in the graph) shows the results for individual room temperature regulation.

In addition, the simulations were conducted both for a building with type A (white bars) and for a building with type B (grey bars) average transmittances.

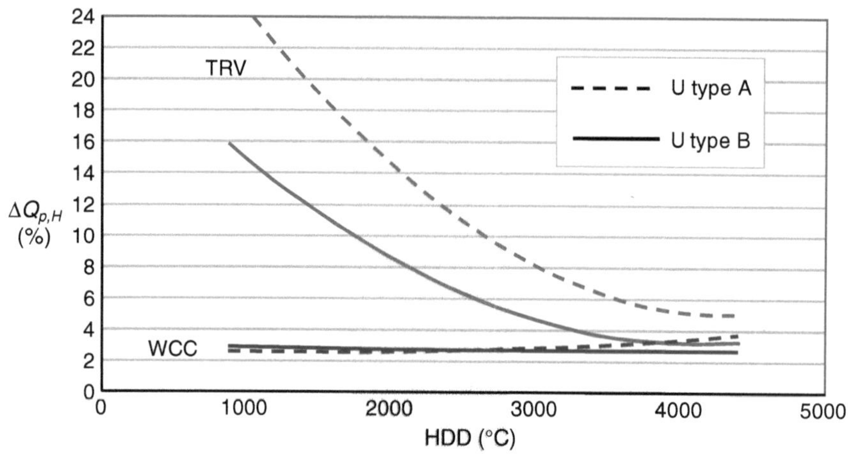

Fig. 11 Radiant panel system. Installation of WCC or TRV. Condensing boiler

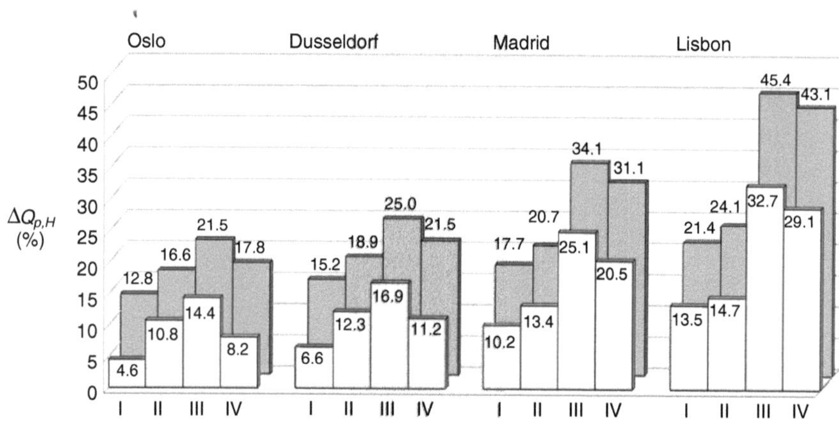

Fig. 12 Replacement of a standard boiler with a condensing boiler. Radiator system

Figure 12 illustrates the results of the simulations which assume, in the as-built state, the presence of a standard-type boiler and a regulation system consisting of a zone thermostat. For the as-planned state, the boiler was assumed to be of the condensing type, with a heat emission system consisting of radiators.

The results show that in addition to the clear energy benefit of replacing the boiler, significant improvements are obtained in each of the climate zones considered. Energy savings increase in warmer climates and are maximised when the condensing boiler is combined with a regulation system which incorporates both weather-based control of the temperature of the water received by the heating elements and air temperature management at the individual room level.

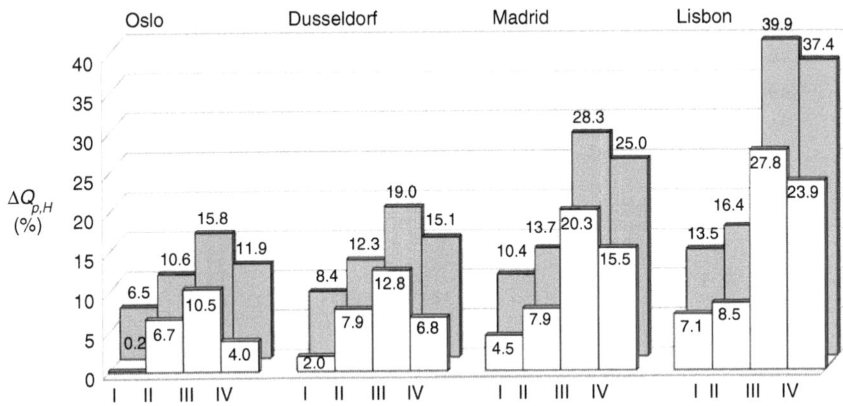

Fig. 13 Replacement of a low-temperature boiler with a condensing boiler. Radiator system

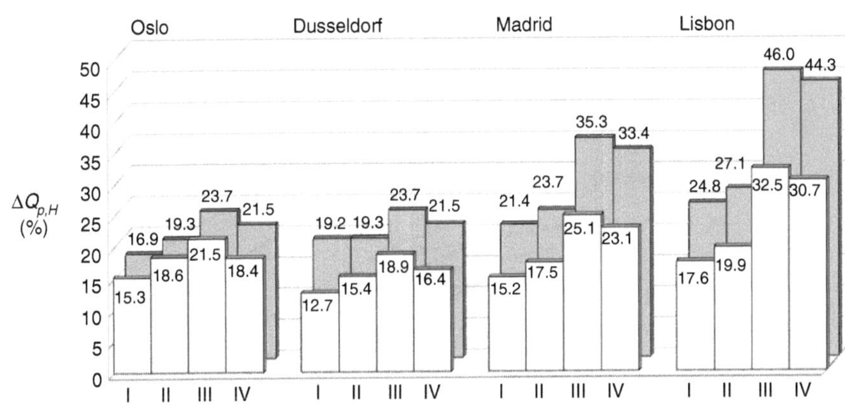

Fig. 14 Replacement of a standard boiler with a condensing boiler. Radiant panel system

Figure 13 shows the results obtained by substituting a low-temperature boiler with a condensing boiler. The results are characterised by patterns similar to the ones in Fig. 12. The percentage values in the second case are lower as a consequence of the fact that the original generator is more efficient than one of the standard type.

The simulations were also conducted considering a low-temperature radiant panel system. The results obtained are illustrated in Figs. 14 and 15.

One of the upgrades simulated was the replacement of a condensing boiler with an electrical air-to-water heat pump. The characteristics of the system were assumed to be such that this replacement could be effected without other types of upgrade that might impact overall performance.

The results obtained highlight the fact that the energy benefit deriving from the implementation of this type of upgrade depends on the nominal output of the heat

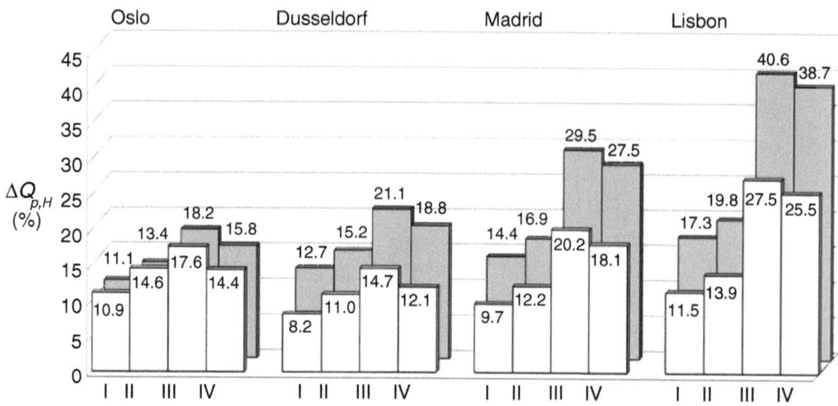

Fig. 15 Replacement of a low-temperature boiler with a condensing boiler. Radiant panel system

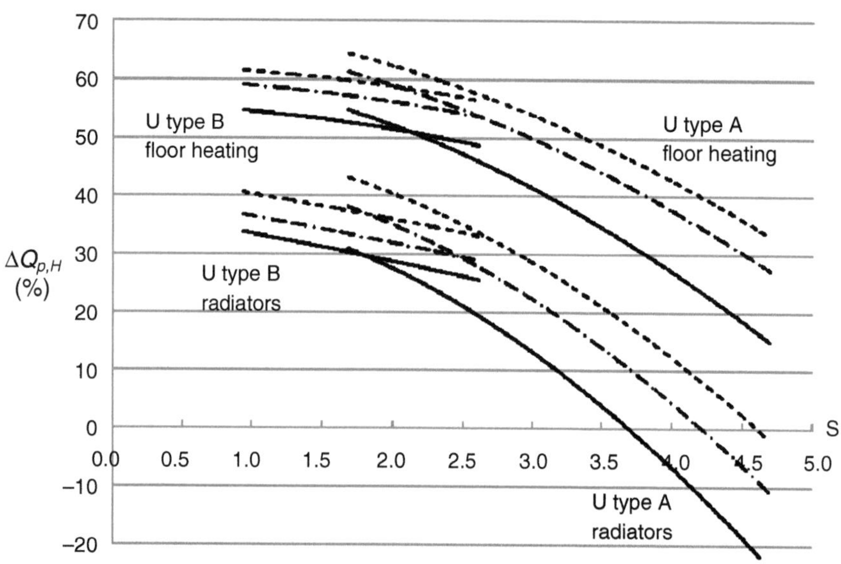

Fig. 16 Comparison between electrical air-to-water heat pump and gas-fired condensing boiler plotted against changes in the oversizing factor *S*. Climate data for the simulation: Northern Europe

pump selected, that is, the oversizing of the machine installed in relation to the heating output required by the system. It was observed that in cases in which the heat pump is not oversized in respect of the building's heat losses, the energy benefit gained from the use of the heat pump as opposed to the combustion boiler is very high. This advantage decreases as the oversizing factor, represented by *S* in Fig. 16, increases.

The simulations were conducted both for radiators and for low-temperature radiant panel systems. Thus, the technical characteristics of the heat pump are, in the first case, those of a device which can produce hot water at a temperature of 75 °C, while in the second case, the heat pump is of a type suitable for producing water at a low temperature.

The results obtained confirm the importance of the correct sizing of the heat pump according to the building's energy requirements.

In these simulations, energy requirements connected with the production of hot water for domestic use were not taken into consideration.

The advantage of installing a heat pump in the place of a condensing boiler remains highly dependent on *S* also with changes in climatic conditions.

It is important to make a number of additional considerations regarding the comparison between as-planned states in which the boiler is gas-fired and that in which the heat source is an electrical heat pump. As it is necessary to make the two forms of energy comparable, the latter can be converted into primary energy by multiplying its value by the conversion factor $f_{\mathrm{p,el}}$. The value of the conversion factor has a significant impact on the results that are obtained in the comparisons previously illustrated in this section. In the specific case, refer to Fig. 16 ($f_{\mathrm{p,el}} = 0.46$).

4.2 Multi-Family Buildings

In order to define the geometrical model of a multi-family building, reference was made to the type illustrated in Fig. 17.

The building is characterised by a layout in the form of an "H" enclosed within a 27 × 27 m square, a surface area of approximately 420 m^2 per floor and a heated volume of around 7,800 m^3. The building has six above-ground floors. Unheated ancillary rooms are located on the first level. The other floors are used for habitation, with six apartments per floor. The core that separates the two main sections is used as a stairwell and is also considered to be an unheated space. The building has a surface-to-volume ratio of 0.46 m^{-1}. Each floor has three types of apartment: the first and second types are characterised by a floor area of approximately 85 and 110 m^2, respectively, while the third type has a floor area of approximately 55 m^2.

It was assumed that the heating system is decentralised without heat metering.

The simulations were conducted both for radiator-based systems and for low-temperature radiant panel systems.

To make them easier to consult and compare, the results have been represented using graphs which are differentiated according to the type of heating element.

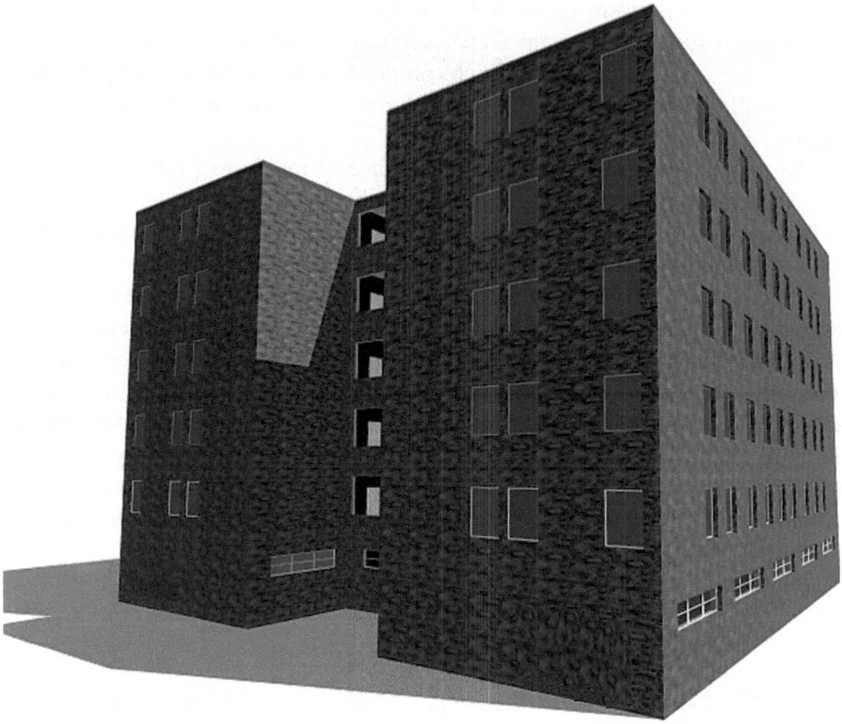

Fig. 17 Model of multi-family building

4.2.1 Upgrades to the Regulation System

With the initial analyses, an assessment was made regarding the degree of impact which the choice of regulation system has on the building's primary energy requirements. In this connection, it was assumed that the as-built state consisted of regulation of the temperature of the water supplied to the heating elements obtained using a weather-based climate control system equipped with an outdoor sensor.

The upgrades considered regard the installation of a temperature control thermostat at the apartment level (ZC in the graphs), or the installation of an individual ambient room temperature control system (RC in the graphs). Regulation of the room-by-room type is achieved by means of thermostatic sensors to be installed on the valves of the heating elements if they are radiators. These installations can be carried out subject to verification of the compatibility of this solution with the characteristics of the circulators of the distribution system.

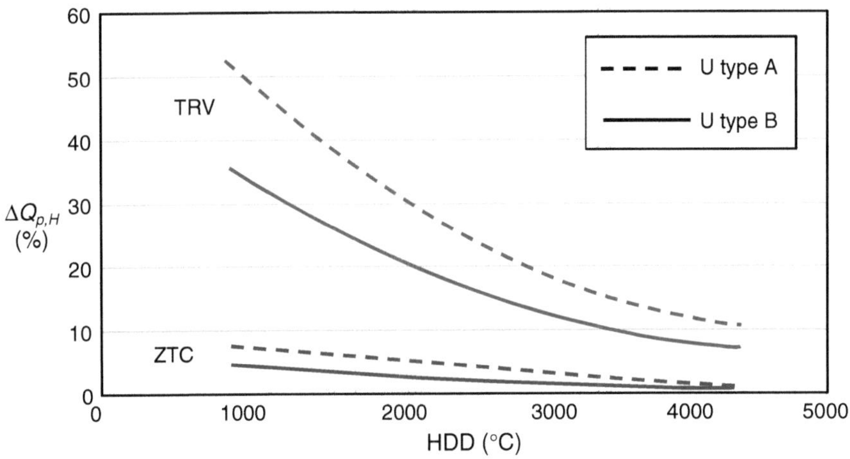

Fig. 18 Radiator system. Standard boiler. Installation of ZC or RC

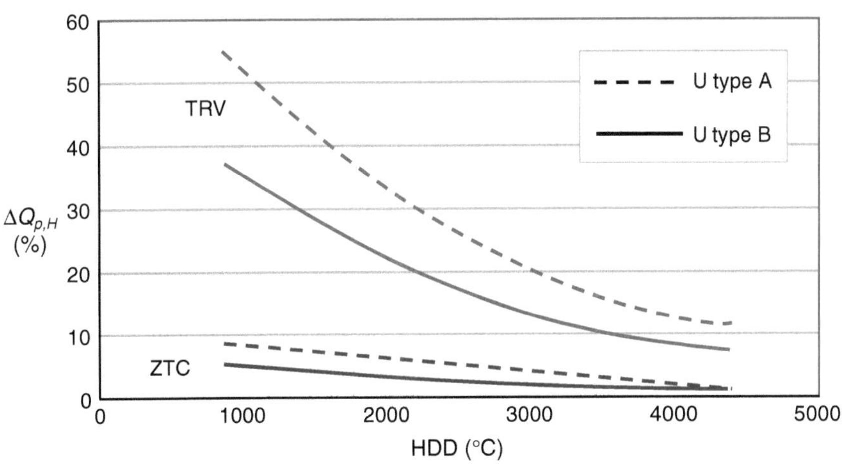

Fig. 19 Radiator system. Low-temperature boiler. Installation of ZC or RC

Figures 18, 19 and 20 refer to cases in which the heating elements are radiators. The heat source considered is, respectively, a standard boiler in Fig. 18, a low-temperature boiler in Fig. 19 and a condensing boiler in Fig. 20.

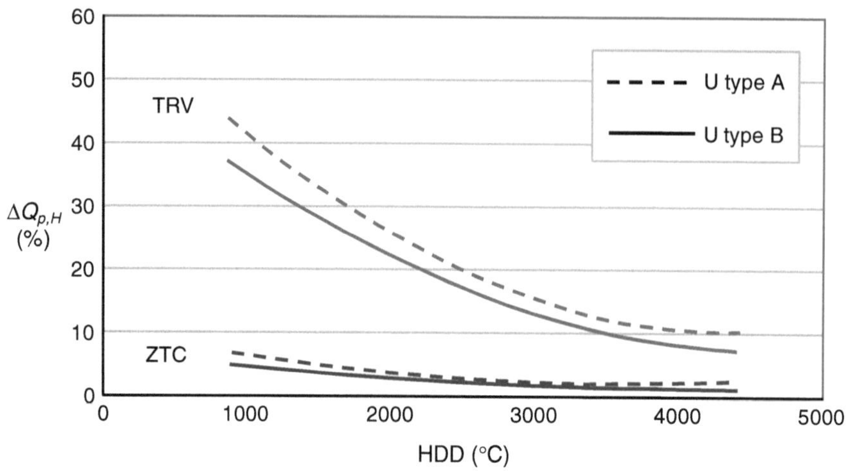

Fig. 20 Radiator system. Condensing boiler. Installation of ZC or RC

In percentage terms, the primary energy saving changes very little with the change in the type of heat source used and is greater for buildings located in warm climates. The largest savings are obtained with the installation of RC systems, due to their more efficient use of free heat gains.

The same type of simulation, with reference to a low-temperature radiant panel system, provided very similar results to the ones just presented.

4.2.2 Upgrades to the Generation System

The purpose of the analysis conducted is to assess the influence on primary energy requirements of a modification to the system of heat energy generation. For each of the cases considered, several simulations were conducted: in the first (I in the graph), it was assumed that no modifications are made to the regulation system; in the second (II in the graph), it was also assumed that one thermostat per apartment is installed, while in the third (III in the graph), it was assumed that individual room temperature regulation control was installed. The simulations were conducted both for a building consisting of components with average transmittance values corresponding to Type A (shown in the graphs in white) and for a building with components with average transmittance values corresponding to Type B (shown in the graphs in grey).

Figure 21 shows the results of the simulations which assume that a standard boiler and a radiator-based system are installed in the as-built state. It was assumed that the generator in the as-planned state is a condensing boiler.

A simulation of the same type, yet which assumes that a low-temperature boiler is installed in the as-built state, produced the results shown in Fig. 22. It will be

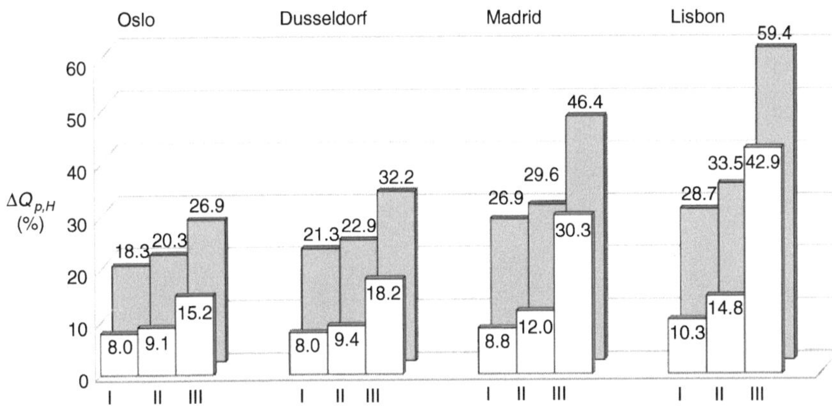

Fig. 21 Replacement of a standard boiler with a condensing boiler. Radiator system

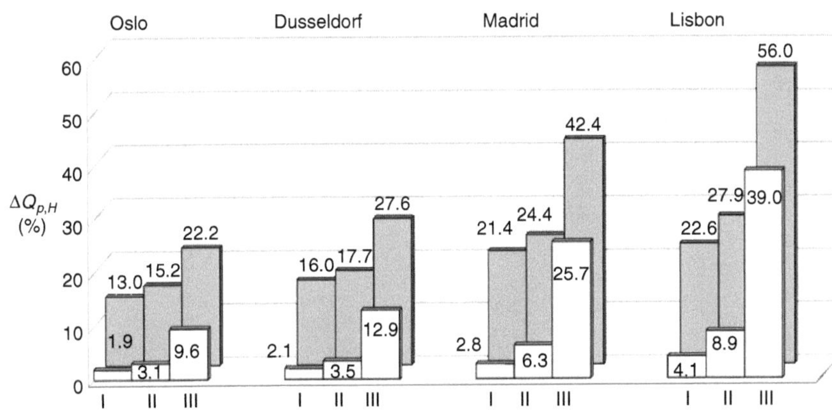

Fig. 22 Replacement of a low-temperature boiler with a condensing boiler. Radiator system

noted that in this case, the energy savings are always lower in percentage terms than those shown by the graph in Fig. 21; this result can be attributed to the fact that in the as-built state in the second case, a higher-efficiency boiler is installed compared to a standard boiler.

The same simulations were also conducted for a radiant panel heating system. The results of these simulations are illustrated by the graphs in Figs. 23 and 24.

As with the simulations conducted on the single-family building, the primary energy saving deriving from the replacement of a condensing boiler with an electrical heat pump of the air-to-water type was assessed. The assessment was conducted exclusively for the case of a heating system with low-temperature radiant panels.

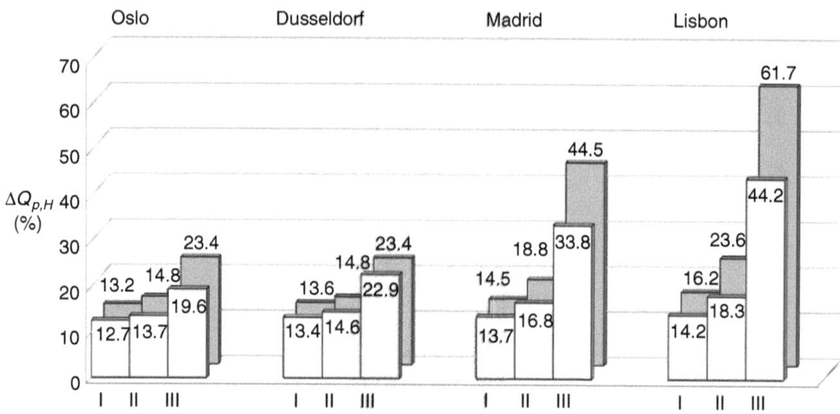

Fig. 23 Replacement of a standard boiler with a condensing boiler. Radiant panel system

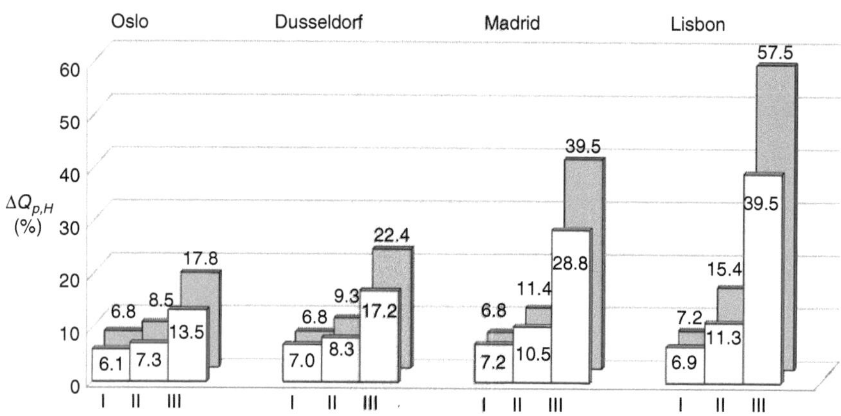

Fig. 24 Replacement of a low-temperature boiler with a condensing boiler. Radiant panel system

From the results obtained, illustrated by the graph in Fig. 25, it can be observed that the primary energy saving is very high in all climate zones, especially if we consider that the boiler considered in the as-built state is a condensing boiler.

The energy saving is higher in the colder climate zone and decreases as one moves towards the warmest climates. This is due to the choice of the heat pump, which is the same for different climate zones and therefore is not oversized with regard to the building losses only in the north. Therefore, the decrease in energy savings obtained in the other climate zones is produced by the effect of the oversizing of the heat pump and mirrors the results already reported for the case of the single-family building.

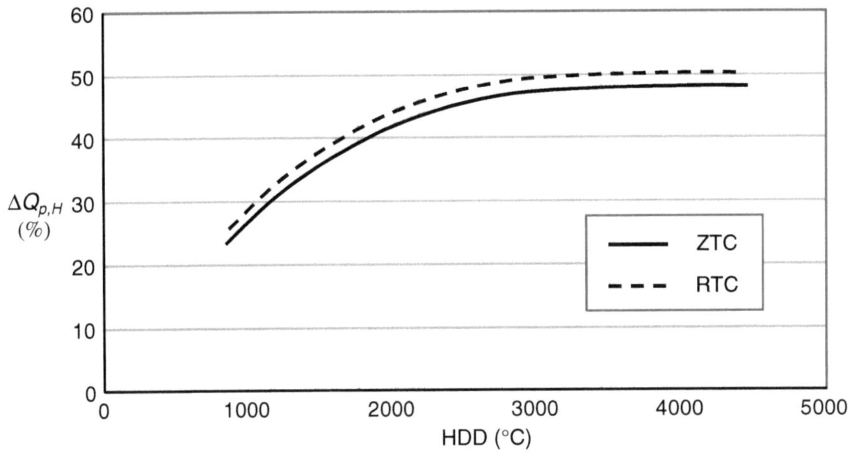

Fig. 25 Replacement of a condensing boiler with an electrical air-to-water heat pump. Radiant panel system

5 Tables for Evaluating Heating System Improvements

Below we present a number of tables which make it possible to estimate the improvement in energy efficiency resulting from a given energy efficiency upgrade. In order to use them, it is necessary to identify the characteristic configuration of the building in question and its heating systems in their as-built state. On the basis of this information, the correct column of the table is selected, while each row corresponds to an as-planned state.

Tables key			
SB	Standard boiler	HT	High-temperature heating system
LTB	Low-temperature boiler	LT	Low-temperature heating system
CB	Gas condensing boiler	ZC	Zone control
HP	Heat pump	WCC	Weather climate control
		RC	Room control
Saving class			
A+	>50 %	D	15–20 %
A	40–50 %	E	10–15 %
B	30–40 %	F	5–10 %
C	20–30 %	G	<5 %

5.1 Single-Family Buildings

North Europe		As built states											
As planned states		SB				LTB				CB			
		HT		LT		HT		LT		HT		LT	
		ZC	ZC WCC	ZC	ZC WCC	ZC	ZC WCC	ZC	ZC WCC	ZC	ZC WCC	ZC	ZC WCC
SB – ZC – WCC	U_a	G		G									
	U_b												
SB – RC – WCC	U_a	F	F		G								
	U_b		G										
LTB – ZC – WCC	U_a					G		F					
	U_b							G					
LTB – RC – WCC	U_a					E	F	E	F				
	U_b					F	G	F	G				
CB - ZC	U_a	E		D				E					
	U_b	G		E		G		F					
CB - ZC – WCC	U_a	D				E	F	E	F	G		G	
	U_b	E	F		D	F	G			F			
CB - RC – WCC	U_a	C				D	E				F		F
	U_b	E			C	E	F	D	E	E	G	F	G
CB - RC	U_a	D				E				F			
	U_b	F			D	G		E		G		G	
HP (S=2)- ZC - WCC	U_a	C				E		D		F			
	U_b												
HP (S=2)- RC - WCC	U_a	C		A		D		A		E		A	
	U_b	D										B	

Central Europe		As built states											
As planned states		SB				LTB				CB			
		HT		LT		HT		LT		HT		LT	
		ZC	ZC WCC	ZC	ZC WCC	ZC	ZC WCC	ZC	ZC WCC	ZC	ZC WCC	ZC	ZC WCC
SB – ZC – WCC	U_a	G		G									
	U_b												
SB – RC – WCC	U_a	E											
	U_b	E	F										
LTB – ZC – WCC	U_a					F		F					
	U_b							G					
LTB – RC – WCC	U_a					C	D	C	E				
	U_b					D	E		F				
CB – ZC	U_a	D		C		E		E					
	U_b	E		D		G		F					
CB - ZC – WCC	U_a	C	D	C		E	F	D	F	G		G	
	U_b	E		D		F	G	E					
CB - RC – WCC	U_a	B				C				D			
	U_b	C					D	C	D		E		
CB – RC	U_a	B		B		C			C	D		D	
	U_b	C		C		D		D	E			F	
HP (S=2)– ZC - WCC	U_a	B		A+		C		A		C		B	
	U_b	C											
HP (S=2)– RC - WCC	U_a	A		A+		B		A+		C		A+	
	U_b	B											

South Europe		As built states											
As planned states		SB				LTB				CB			
		HT		LT		HT		LT		HT		LT	
		ZC	ZC WCC	ZC	ZC WCC	ZC	ZC WCC	ZC	ZC WCC	ZC	ZC WCC	ZC	ZC WCC
SB – ZC – WCC	U_a	G		G									
	U_b												
SB – RC – WCC	U_a	C											
	U_b	D		E									
LTB – ZC – WCC	U_a					F		E					
	U_b							F					
LTB – RC – WCC	U_a					B	C	B	C				
	U_b					C	D	C	D				
CB - ZC	U_a	C		C		E		D					
	U_b	E		D		F		E					
CB - ZC – WCC	U_a	C				D	F	D	F	G		G	
	U_b	E		D		F	G	E					
CB - RC – WCC	U_a	A				B		A	B		C		
	U_b	B				C						D	
CB - RC	U_a	A		A		B		A	B	C		C	
	U_b	C		B		C		C		D		D	
HP (S=2) – ZC – WCC	U_a	A		A+		A				B			
	U_b												
HP (S=2) – RC – WCC	U_a	A+		A+		A		A+		A		A+	
	U_b	A								B			

5.2 Multi-Family Buildings

North Europe		As built states											
As planned states		SB				LTB				CB			
		HT		LT		HT		LT		HT		LT	
		ZC	ZC WCC	ZC	ZC WCC	ZC	ZC WCC	ZC	ZC WCC	ZC	ZC WCC	ZC	ZC WCC
SB – ZC – WCC	U_a	G		G									
	U_b												
SB – RC – WCC	U_a	F	F		G								
	U_b		G										
LTB – ZC – WCC	U_a					G		F					
	U_b							G					
LTB – RC – WCC	U_a					E	F	E	F				
	U_b					F	G	F	G				
CB - ZC	U_a	E		D				E					
	U_b	G		E		G		F					
CB - ZC – WCC	U_a	D				E	F	E	F	G		G	
	U_b	E	F	D		F	G		E	F			
CB - RC – WCC	U_a	C				D	E	D			F	F	F
	U_b	E		C		E	F			E	G		G
CB - RC	U_a	D				E				F			
	U_b	F		D		G		E		G		G	
HP (S=2) – ZC – WCC	U_a	C				E		D		F			
	U_b												
HP (S=2) – RC – WCC	U_a	C		A		D		A		E		A	
	U_b	D										B	

Central Europe		As built states											
As planned states		SB				LTB				CB			
		HT		LT		HT		LT		HT		LT	
		ZC	ZC WCC	ZC	ZC WCC	ZC	ZC WCC	ZC	ZC WCC	ZC	ZC WCC	ZC	ZC WCC
SB – ZC – WCC	U_a	G		G									
	U_b												
SB – RC – WCC	U_a	E											
	U_b	E	F										
LTB – ZC – WCC	U_a					F		F					
	U_b							G					
LTB – RC – WCC	U_a					C	D	C	E				
	U_b					D	E		F				
CB - ZC	U_a	D		C		E		E					
	U_b	E		D		G		F					
CB - ZC – WCC	U_a	C	D	C		E	F	D	F	G		G	
	U_b	E		D		F	G	E					
CB - RC – WCC	U_a	B				C				D			
	U_b	C					D	C	D		E		
CB - RC	U_a	B		B		C			C	D		D	
	U_b	C		C		D		D	E			F	
HP (S=2) – ZC – WCC	U_a	B		A+		C		A		C		B	
	U_b	C											
HP (S=2) – RC – WCC	U_a	A		A+		B		A+		C		A+	
	U_b	B											

South Europe		As built states											
As planned states		SB				LTB				CB			
		HT		LT		HT		LT		HT		LT	
		ZC	ZC WCC	ZC	ZC WCC	ZC	ZC WCC	ZC	ZC WCC	ZC	ZC WCC	ZC	ZC WCC
SB – ZC – WCC	U_a	G		G									
	U_b												
SB – RC – WCC	U_a	C											
	U_b	D		E									
LTB – ZC – WCC	U_a					F		E					
	U_b							F					
LTB – RC – WCC	U_a					B	C	B	C				
	U_b					C	D	C	D				
CB - ZC	U_a	C		C		E		D					
	U_b	E		D		F		E					
CB - ZC – WCC	U_a	C				D	F	D	F	G		G	
	U_b	E		D		F	G	E					
CB - RC – WCC	U_a	A				B		A	B		C		
	U_b	B				C						D	
CB - RC	U_a	A		A		B		A	B	C		C	
	U_b	C		B		C		C		D		D	
HP (S=2) – ZC – WCC	U_a	A		A+		A				B			
	U_b												
HP (S=2) – RC – WCC	U_a	A+		A+		A		A+		A		A+	
	U_b	A								B			

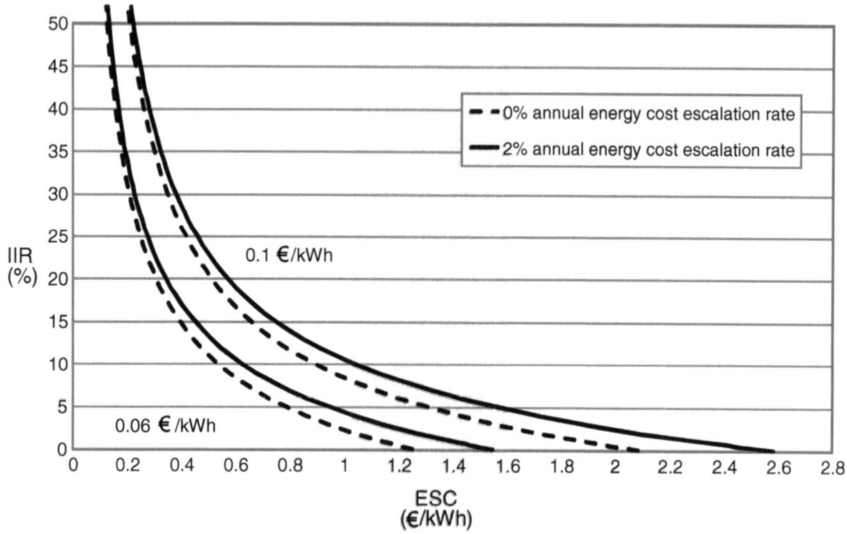

Fig. 26 Function IIR in response to changes in ESC for two different values of C_E and in two different hypothetical values of r

6 Economic Analysis of Energy Efficiency Upgrades to Heating Systems

In order to evaluate the cost-effectiveness of the upgrades in question, instruments usually adopted to analyse the economic viability of a financial investment were used, in particular the internal rate of return (IIR), defined as the discount rate which makes the net present value of a series of cash flows equal to zero. The IIR is given by solving the following equation, (16):

$$\sum_{t=0}^{n} \frac{\mathrm{CF}_t}{(1+i)^t} = 0 \tag{17}$$

An investment project is desirable if the IIR is higher than the tax of another comparable investment.

In our case, it was assumed that the upgrade entails a single negative cash flow corresponding to the initial investment, CF_0, necessary to carry out the energy efficiency upgrade and that each successive financial flow, CF_t, consists of the product of the annual energy saving, which is assumed to be constant over time, and the cost per unit of primary energy. Therefore, Eq. (16) can be rewritten in the following form:

$$\mathrm{CF}_0 = \sum_{t=1}^{n} \frac{\mathrm{CF}_t}{(1+i)^t} \tag{18}$$

In order to obtain a function that is independent of the type of upgrade considered as well as of the climatic conditions of the place in which it is carried out, a variable called energy-saving cost (ESC) has been introduced. It is defined as follows:

$$\mathrm{ESC} = \frac{\mathrm{CF}_0}{\Delta Q_{\mathrm{p,H}}} \tag{19}$$

Following the introduction of this variable, Eq. (17) can be restated so that we obtain.

$$\mathrm{ESC} = C_{\mathrm{E}} \sum_{t=0}^{n} (1+r)^t \frac{1}{(1+i)^t} \tag{20}$$

The graph corresponding to Eq. (19) can be represented with parametric curves against changes in cost CE. An example is given in Fig. 26.

In order to illustrate the use of the graph in Fig. 26, an evaluation of three alternative upgrades—named, for the sake of simplicity, Type 1, Type 2 and Type 3. The upgrades concern a condominium-type building with average transmittance value of the envelope corresponding to Type B in Table 3. The Type 1 upgrade consists in the replacement of a standard boiler with a condensation boiler; the Type 2 upgrade involves the installation of a regulation system with thermostatic sensors on a radiator-based system with a standard boiler and zone thermostat; the Type 3 upgrade involves the installation of a weather-compensated control system on a radiator-based system with a standard boiler and zone thermostat. For each of these upgrades, the initial investment and energy saving obtained were estimated. For this purpose, a calculation of the works required to execute each one of the energy efficiency upgrades was made and the graphs in Sect. 5.2 were used to determine primary energy-saving values. Finally, the cost of one unit of primary energy saved, ESC, was calculated using Eq. 3.19. Table 4 contains the reference values for the three alternatives considered, evaluated under two different climatic conditions.

The results obtained allow us to affirm that the operation of substituting the heat generator yields the greatest benefit in cold climates, yet its cost-effectiveness needs to be evaluated carefully if compared with the installation of an individual room temperature control system. In this connection, it is pointed out that the cost of the latter type of upgrade is more sensitive to the number of rooms in the building in relation to the cost to be borne in replacing the heat generator. The evaluation leads to different results in a warm climate characterised by greater solar heat gains: in such a case, in the example proposed, the most cost-effective energy efficiency upgrade consists in the installation of individual room temperature control.

In general, in order to use the graph in Fig. 26, it is necessary to know the primary energy requirements of the existing building in its as-built state. Using the tables in Sect. 6, or the graphs in Sect. 5, it is possible to evaluate the potential

Table 4 Evaluation of a number of hypothetical energy efficiency upgrades to heating systems on a centralised building (C_E = 0.1€/kWh; r = 0.02)

	Type 1 (CF_0 = 18,000€)			Type 2 (CF_0 = 19,000€)			Type 2 (CF_0 = 14,000€)		
	$\Delta Q_{p,H}$ kWh	ESC €/kWh	IIR %	$\Delta Q_{p,H}$ kWh	ESC €/kWh	IIR %	$\Delta Q_{p,H}$ kWh	ESC €/kWh	IIR %
Oslo	34,311	0.5	23	33,896	0.6	19	5,348	2.6	0.01
Lisbona	12,144	1.5	5	41,605	0.5	23	5,865	2.4	0.7

energy saving that can be achieved with the upgrade. By estimating the cost to be borne in order to effect the upgrade and dividing this value by the amount of energy saved expressed in kilowatt hours, it is possible to determine the parameter ESC, using which in the graph in Fig. 26 it is possible to estimate the value of the internal rate of return on the investment under consideration.

References

1. Ceccotti L, De Angelis A, Saro O (2008) Modulation effect on energy efficiency of condensing boilers (in Italian). 46° International Conference AICARR, 12–13 March 2008
2. Cernotta F, Cortella G, De Angelis A, Saro O Energy refurbishment of one-family buildings by retrofitting of heating plants (in Italian). La Termotecnica, n.05/11
3. Ceccotti L, De Angelis A, Saro O (2007) Evaluation of energy saving in thermal plants with modulating boilers coupled to radiators (in Italian). 62° National Conference ATI, 11–14 Sept 2007

Standard References

4. ASHRAE Standard Committee, ASHRAE HANDBOOK: Fundamentals 2013, ASHRAE
5. CEN, Heating systems in buildings—Method for calculation of system energy requirements and system efficiencies—Part 4: Space heating generation systems, combustion systems (boilers) EN 15316-4-1:2008
6. European Union, Directive 92/42/EEC of the European Parliament and of the Council of May 21th, 1992 on efficiency requirements for new hot-water boilers fired with liquid or gaseous fuels. Official Journal of the European Union, 21 May, 1992
7. ISO, Energy performance of buildings—Calculation of energy use for space heating and cooling ISO 13790: 2008
8. UNI, Technical Specification, Energy performance of buildings. Part 1. Evaluation of energy need for space heating and cooling, UNI/TS 11300-1, 2008
9. UNI, Technical Specification, Energy performance of buildings. Part 2: Evaluation of primary energy need and of system efficiencies for space heating and domestic hot water production, UNI/TS 11300-2, 2008

Solar Energy

G. Oliveti, L. Marletta, N. Arcuri, M. De Simone, R. Bruno and G. Evola

Abstract European Directive 2010/31/CE on energy efficiency in the buildings sector provides for significant actions for the reductions in energy consumption, and Directive RES 2009/28/CE stimulates the use of energy from renewable sources in order to meet such objectives. The chapter presents indications about the use of solar radiation for the energy requalification of buildings based on the results of research activities. Simplified evaluation methods are presented with the aim to verify the available potential energy, for the production of sanitary hot water, for winter heating and for the production of electrical energy, by means of systems, which use conventional solar collectors placed on the surfaces of the building shell, in particular on the roof slopes. In order to evaluate the energy improvement linked to the solar gain through the windows, the direct gain is evaluated by means of an accurate calculation model of the solar gains, which uses the coefficient of effective absorption of the entering radiation. With regard to sunspaces, some aspects of the thermal analysis, of the evaluation of the solar energy absorbed by the sunspace and by the adjacent room and of the benefits obtainable in terms of a reduction in the thermal requirements of the adjacent spaces are discussed. Finally, a discussion is presented regarding the possibility of using phase change materials (PCM) for the refurbishment of lightweight buildings. This technique allows for the improvement of the response of the building to solar gains, thus providing better thermal comfort in summer. In order to facilitate comprehension, the topics are supported by calculation methods and accompanied by numerical examples.

G. Oliveti · N. Arcuri (✉) · M. De Simone · R. Bruno
Department of Mechanical, Energy and Management Engineering,
University of Calabria, Cosenza, Italy
e-mail: natale.arcuri@unical.it

L. Marletta · G. Evola
Department of Industrial Engineering, University of Catania, Catania, Italy

A. Magrini (ed.), *Building Refurbishment for Energy Performance*,
Green Energy and Technology, DOI: 10.1007/978-3-319-03074-6_4,

Nomenclature

A_c Collection surface area (m^2)
a_w Azimuth (°)
C_{eq} Equivalent specific heat capacity of PCM (J/kg K)
c_p Specific heat of the air (J/kg K)
$\bar{E}_{ass}$ Daily average monthly solar energy absorbed by an internal space (J)
E_i Daily average monthly energy incident on the external glazed surface (J/m^2)
f Glazed fraction of a wall (-)
f_c Corrective factor (-)
F_R Removal factor of the thermal collector (-)
F' Efficiency factor of the thermal collector (-)
g Total solar gain for normal incidence of the glazed system (-)
G_c Solar global irradiation on a surface (W/m^2)
G_e Solar power entering in an environment through the glazed surface (W/m^2)
h_c Convective thermal exchange coefficient (W/m^2 K)
I_{bo} Direct solar irradiation on the horizontal plane (W/m^2)
I_{do} Diffuse solar irradiation on the horizontal plane (W/m^2)
$\dot{m}_v$ Ventilation flow rate (kg/s)
P_{cel} Electrical power supplied by the PV cell (W)
Q_{ai}^{+} Heat transferred by convention to the internal air (J)
Q_i Incident solar energy on the sunspace shell (J)
Q_{ass} Solar power absorbed by the internal environment, or net solar gain (W)
$Q_{as,s}$ Solar energy absorbed in the sunspace (J)
Q_p Lost thermal power of solar collector (W)
Q_{sol} Daily average monthly solar gain through the glazed surfaces (J)
Q_{tr} Transmitted solar energy through the sunspace shell (J)
Q_u Useful thermal power of solar collector (W)
R_b Inclination factor of direct solar radiation (-)
$\bar{R}_b$ Monthly direct radiation inclination factor (-)
R_d Inclination factor of diffuse solar radiation (-)
R_r Inclination factor of reflected solar radiation (-)
T_a Outdoor air temperature (K)
T_{as} Air temperature in the sunspace (K)
T_c Average temperature of the PV cell or panel (K)
T_i Internal surface temperature (K)
T_p Peak melting temperature of PCM (K)
$\bar{T}_p$ Average temperature of the thermal solar collector absorbent plate (K)
U Thermal transmittance (W/m^2 K)

Greek Symbols

α Absorption coefficient of solar radiation (-)
α_{cav} Effective absorption coefficient of the internal environment (-)

α_i	Absorption coefficient in the solar band of the *i*th surface (-)
α_f	Solar absorption coefficient of the floor and of the walls (-)
α_m	Average absorption coefficient of the opaque surfaces of the internal environment (-)
α_s	Effective absorption coefficient of the sunspace (-)
α_w	Solar absorption coefficient of the walls (-)
β	Inclination (°)
Δt	Time interval (s)
η	Efficiency (-)
η_u	Utilisation factor (-)
$\eta_{u,v}$	Utilisation factor imputable to ventilation (-)
τ	Transmission coefficient of solar radiation (-)
τ_b	Transmission coefficient of the direct solar radiation of the glazed system (-)
τ_d	Transmission coefficient of the diffuse solar radiation of the glazed system (-)
τ_g	Transmission coefficient of the reflected solar radiation of the glazed system (-)
ψ	Glazed fraction of the room (-)

1 General Considerations of the Use of Solar Thermal Energy

In energy requalification interventions on buildings, solar energy can be used with active systems in which the transport of the energy collected is carried out with fluids in movement under the action of pumps or fans, or even the use of passive systems, in which collection and storage can be combined in a sole component and energy transfer is entrusted to thermal irradiation and to transport due to natural convention, without the contribution of auxiliary energy.

In active systems, solar radiation is used for the heating of a thermo-vector fluid by means of a particular heat exchanger which is the solar collector. Such a component transforms radiant energy into thermal energy which is then used for many different aims, but mainly for winter heating and for the production of domestic hot water (DHW).

Besides the solar collector, due to the precariousness and discontinuity of the solar source, these systems are equipped with one or more storage tanks with water, which store the surplus thermal energy in order to return it, upon the user's request, at different periods.

With water plants, the heating of environments requires the use of low-temperature emission terminals, such as radiant panels or fan coils, in which the efficiency of solar collection is much higher when the required temperature for the thermo-vector fluid is lower.

In order for solar plants to be financially convenient, they must be dimensioned in such a way as to supply only a fraction of the energy thermal requirement, while the remaining part is supplied by an auxiliary system. The correct sizing of such plants requires economic evaluation and optimisation methods of the system formed by the solar plant and by the integration system. The main project parameter is represented by the area of collectors A_c: with an increase in the collection area, the collected energy and energy saving increases, but, at the same time, the cost of the plant increases. In order to realise a solar plant, high investment costs are required, which must be addressed prior to starting the plant, and low operating costs which are renewed each year. Economic evaluations are obtained through an analysis of the costs and the gains for the entire life of the plant, updating expenditure and income over several years.

The optimal dimension of the investment can be obtained by determining the value of the collection area which renders the maximum net present value (NPV), or by means of optimisation methods such as the global cost method and the global saving method. The economic variable, such as the global cost of the plant or the global saving obtained, defined as the difference between the global cost of the conventional plant which uses usual sources and the global cost of an integrated solar plant, is expressed as the function of the area of collectors. Such an area is determined in such a way as to maximise or minimise the chosen economic function. In the global cost method of an integrated solar system, the cost relating to the conventional integration plant is not considered in which it is supposed that the latter must be present in every case.

2 Thermal Analysis of Solar Collectors

Thermal solar collectors are simple devices. The flat type, which is the most common and the most economic, is formed by a radiation collection plate, by one or more glass coverings, in order to reduce thermal loss externally, as well as a system of channels connected to the plate through which a thermo-vector fluid flows to remove power. A containment box completes the structure in which the rear and lateral insulation of the panel is inserted.

The instantaneous thermal balance equation of the absorbent plate in stationary regime conditions can be expressed as

$$G_c A_c(\tau\alpha) = Q_u + Q_p \tag{1}$$

with G_c global irradiation incident on the collector (W/m^2), $(\tau\alpha)$ effective product of the transmission coefficient of the glazed covering system and the absorption coefficient of the absorbent black plate, A_c collection surface area, Q_u useful power delivered by the plate to the thermo-vector fluid and Q_p thermal power lost due to convection and irradiation by the collector to the external environment.

The power lost is evaluated with the relation:

$$Q_p = U_c A_c(\bar{T}_p - T_a) \tag{2}$$

with U_c global thermal exchange coefficient between the plate and the air, $\bar{T}_p$ average temperature of the absorbent plate and T_a outdoor air temperature.

Through relations (1) and (2), the useful power can be expressed as the difference between the power absorbed and the power lost:

$$Q_u = G_c A_C \, \tau\alpha - U_c A_C \left(\bar{T}_p - T_a\right) \tag{3}$$

Efficiency is the parameter used to thermally qualify the collector. The instantaneous efficiency η of the collector is defined as the relation between the useful power and the incident solar power:

$$\eta = \frac{Q_u}{G_c A_c} \tag{4}$$

The average efficiency in a time interval (an hour, a day, a month) is defined by means of the equation:

$$\eta = \frac{\int_t^{t+\Delta t} Q_u(t)\mathrm{d}t}{\int_t^{t+\Delta t} A_c G_c \mathrm{d}t} \tag{5}$$

The expression (3) does not allow for the calculation of the useful power Q_u, given that the average temperature of the absorbent plate is not known. The plate is the seat of a bidirectional temperature distribution, in the flow rate direction within the channels and in a perpendicular direction; the determination of which allows the calculation of the power that is transferred by the plate to the cooling liquid.

In the thermal analysis of the solar collectors, the collector efficiency factor F' and the heat removal factor F_R are defined [1, 2]. The physical meaning of the former factors is the following: F' represents the relation between the thermal resistance between the absorbent plate and the external environment, in the hypothesis of uniform plate temperature, and the thermal resistance between the fluid flowing in the channels and the external environment, F_R is the ratio between the power taken by the cooling flow rate and the power that should be taken by it were the plate to have uniform temperature and equal to the inlet temperature of the fluid. This power is the maximum transferable to the fluid since in the considered conditions, the power lost is minimal.

The introduction of factors F' and F_R allows for the expression of the useful thermal power transferred to the flow rate with the relations:

$$Q_u = F' A_c \left[(\tau\alpha) G_c - U_c \left(\bar{T}_f - T_a\right)\right] \tag{6}$$

$$Q_u = F_R A_c \left[(\tau\alpha) G_c - U_c \left(T_{fi} - T_a\right)\right] \tag{7}$$

Equation (7) consents the direct calculation of the useful power, given that the inlet temperature of the fluid is commonly known, unlike (6) in which the average temperature appears, which is generally not known, given that the outlet temperature of the water flow rate is not known. For such a reason, in this case, the resolution is obtained by successive iteration: the editing temperature is set and from (6) the useful power is obtained, and by means of the water flow rate heating equation:

$$Q_u = \dot{m} c_p \left(T_{fu} - T_{fi}\right) \tag{8}$$

the new outlet temperature is obtained.

Table 1 Typical coefficients of the efficiency curve for different types of thermal solar collectors

Collector type	$F'(\tau\alpha)$	$F'U_{co}$ (W/m^2 K)	$k{\cdot}F'$ (W/m^2 K^2)
Unglazed	0.770	9.215	0.700
Flat with one glass covering	0.810	4.360	0.650
Flat with two glass coverings	0.755	2.725	0.095
Selective with one glass	0.840	3.550	0.500
Evacuated tube	0.641	1.059	0.004
Evacuated heat pipe	0.765	0.390	0.002

3 Efficiency Curves of Solar Collectors

Commonly used solar collectors are of different types, and the choice of the most suitable model to be used is dependent, above all, on the required temperature levels, on the thermal energy requirements and also on the period of use during the year [3]. The main collectors available on the market are as follows: unglazed solar collectors, glazed collectors, glazed collectors with selective surfaces, evacuated tube collectors and evacuated heat pipe solar collectors.

If one considers the expression of efficiency (6), in which the average temperature of the fluid between entering and exiting the collector appears, and the most realistic hypothesis is adopted of the coefficient of loss of the collector is not constant but varies in a linear manner in relation to temperature differences:

$$U_{c} = U_{co} + k(\bar{T}_{f} - T_{a}) \tag{9}$$

k is a constant value, the expression becomes

$$\eta = F'(\tau\alpha) - F'U_{co}\frac{\bar{T}_{f} - T_{a}}{G_{c}} - k \cdot F'\frac{(\bar{T}_{f} - T_{a})^{2}}{G_{c}} \tag{10}$$

In Table 1, the values assumed by the coefficients $F'(\tau\ \alpha)$, $F'U_{co}$, and $k{\cdot}F'$ are reported for the main types of collectors [28].

4 Solar Plants

4.1 Introduction

These plants, due to reasons of technical–economical feasibility, are sized to cover a fraction of the energy thermal requirement. From that, it is necessary to equip them with a traditional integration system (boiler or electrical resistance) in order to cover the remaining requirement fraction. Furthermore, due to climatic variability, a water storage tank is always present [4]. For the production of DHW, the most simple plant is that with natural circulation, characterised by the placing of the tank in a higher position compared to the collector.

If it is not possible to place the storage tank higher than the collector, then it will be necessary to resort to forced circulation. A pump placed on the cold branch of the circuit provides for the circulation of the fluid between the storage system and the collector. Forced circulation plants can be equipped with two storage tanks, the first for "preheating" at more limited temperatures in order to improve collection efficiency and the second for "use" interfaced to the auxiliary system.

The installation of solar plants in condominium buildings is more convenient compared to their installation in single residential units. The collectors can be placed on roof slopes, and integration with the existing plant does not usually present any particular difficulties.

These plants can either use natural or forced circulation. In the first case, it is necessary to install monoblock systems in which each panel supplies DHW to a single residential unit. Many gas boilers used for domestic heating and for the production of DHW are semi-modulating and supply a minimum thermal power of 4–5 kW, independently of the inlet water flow rate temperature. Such a power can result as being excessive to obtain the requested outlet temperature of 45 °C. In order to remedy such an inconvenience, it is good practice to install a mixing valve before the boiler which reduces the hot water flow rate taken from the tank and, by means of a cold water flow rate originating from the aqueduct network, reduces the temperature entering the boiler.

In the case in which condensation boilers are used, the supply of minimum power leads to higher generation efficiency due to the use of the heat from condensation present in fumes.

The centralised production of DHW by means of forced circulation plants equipped with a sole service storage system for all the condominium users results as being more rational. This solution is capable of producing significant fractions of the DHW thermal requirement and is easy to integrate with existent plants since it requires a hot water distribution system which is realisable with a single pipe to which all users can be connected.

In water plants for the heating of buildings, the most efficient use of solar radiation is obtained using a low-temperature water flow rate (40–50 °C) with suitable energy distribution systems for the spaces to be heated, such as radiant floors, radiant ceilings, or fan-coil-type devices. The solar plant and the integration system can be configured differently compared to the environment to be heated: the solar plant and the boiler are placed in parallel, or in series, the solar panels and the integration system jointly supply the storage system, or even the solar and the auxiliary are independent with a double distribution system of heat to the spaces. The latter configuration lends itself to being used in buildings that are already equipped with traditional heating plants with radiators.

4.2 Planning Methods of Solar Plants

In order to plan solar plants in an accurate manner, it is necessary to simulate the plant components, such as the array of collectors, the storage system, the heat exchangers, the distribution network, the distribution terminals and the control system, with dynamic models. The equations to consider are those of conservation of the mass and energy for the various plant components.

Often simplified models based on energetic type evaluations whose level of accuracy depends on the entity of simplification are used. For example, the average annual method [3] neglects the dynamic interaction between the storage tank, the field of collectors and the heat use system and for such reasons provides preliminary evaluations. More accurate results are obtained with the f-chart method [2] created based on the results of numerous dynamic simulations of reference solar plants.

Several *f*-chart versions are available: for forced circulation liquid plants for the heating of spaces and production of DHW; for plants which are solely for the production of DHW; for natural circulation liquid plants and also for air plants. Such a method allows for the calculation of the fraction f of the monthly thermal requirement obtained from the solar radiation in a plant with determined properties and the annual solar fraction F. It is a verification method of plant performance which, if opportunely used, allows for the obtainment of an optimal plant design.

Hereafter, two simplified calculation procedures are described. The first regards DHW production plants and the second plants for heating, both regarding existing buildings. The two procedures use roof-covering slopes as sunlight surface, which are differently tilted and oriented, and evacuated heat pipe solar collectors.

4.3 A Simplified Method for the Determination of the Solar Collector Surface of Buildings for the Production of DHW with Evacuated Heat Pipe Collectors

In new buildings, and in those subject to important restructuring work, it is good practice to expect that not less than 50 % of the annual thermal energy requirement for DHW is produced by solar collectors.

The considered method [5] provides for the use of evacuated heat pipe collectors. If used to produce a water flow rate at a temperature of 35–50 °C, these collectors' deliver performance is mainly dependent on solar radiation, while the influence of the temperature difference between the thermo-vector fluid and the external air can be held to be negligible. In such conditions, collectors' efficiency is independent from the storage temperature and therefore from the thermal energy removing from such a component.

The estimation of producible thermal energy of DHW plants (kWh/m^2), with panels that are arbitrarily oriented and tilted, can be determined with a simplified

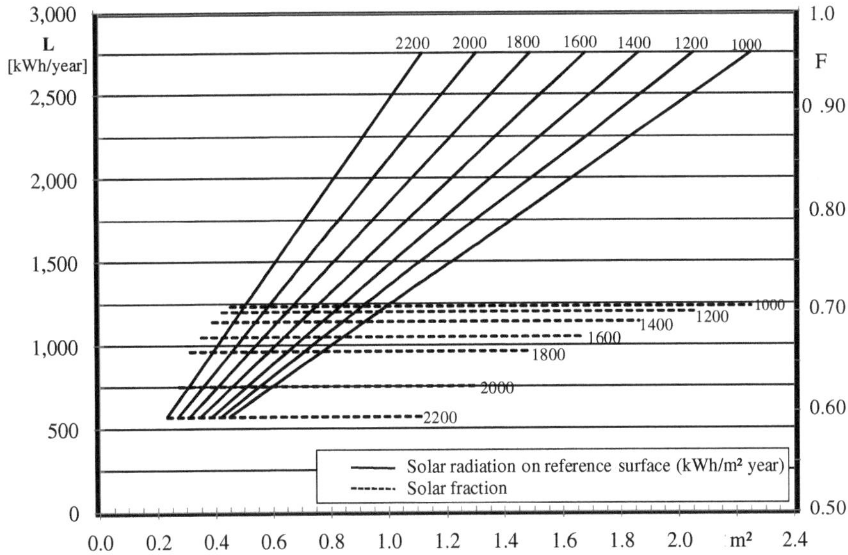

Fig. 1 Collection area and solar fraction as a function of the annual thermal energy requirement for DHW and of the annual solar energy on the reference surface ($\beta = 30°$, $a_w = 0°$). Small residential users

procedure which uses results obtained in reference plants with south-facing collection surfaces ($a_w = 0°$) and tilted by 30° ($\beta = 30°$).

Such an arrangement generally provides the best annual collection conditions for latitudes between 35°N and 45°N.

If the slopes of the roof covering of a building are used as the collection surface, which are generally differently oriented and tilted, the method assigns each slope with a thermal producibility value determined from the value relative to reference setting. In such a way, it is possible to define the best positioning of the collection surface and the area necessary to obtain the same thermal requirement fraction.

The data required by the calculation procedure are the DHW annual thermal energy requirement (kWh/year) and the annual solar energy (kWh/(m^2 year)) incident on a unitary reference surface ($\beta = 30°$; $a_w = 0°$).

Through these data and the graphs in Figs. 1 and 2, it is possible to obtain the collection area and the annual fraction of the thermal requirement provided by the solar source. The annual energy values on the reference plane are variable from 1,000 to 2,200 kWh/(m^2 year) corresponding to the geographic area comprised between the Northern Africa and the northern Europe. Figure 1 relates to small domestic users with annual thermal requirements of less than 3,000 kWh/year, while Fig. 2 relates to larger thermal requirements, though not greater than 45,000 kWh/year.

In order to calculate the storage volume, if L is the DHW annual thermal requirement, the following relation is used (L in kWh and V in litres):

$$V = 0.068\,\mathrm{L} \qquad (11)$$

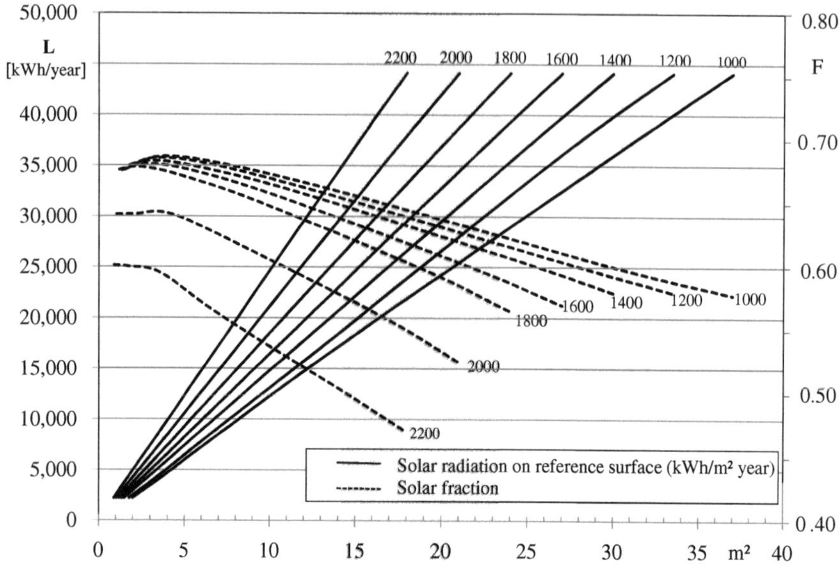

Fig. 2 Collection area and annual solar fraction as a function of the thermal energy requirement for DHW and of the radiation on the reference surface ($\beta = 30°$, $a_w = 0°$). Large residential users

For example, for an annual thermal requirement of 2,500 kWh and an availability of solar radiation on the reference surface of 1,600 kWh/m^2, from Fig. 1, a collection surface of 1.2 m^2 is obtained and an annual solar fraction of the DHW thermal requirement of 0.62 is obtained. Annual thermal producibility is equal to $0.62 \cdot 2{,}500/1.2 = 1{,}292$ kWh/(m^2 year).

For collection surfaces which are differently oriented and tilted, the calculation procedure provides coefficients of reduction in the annual producibility relative to the reference layout. Table 2 shows the corrective coefficients relating to 42°N of latitude for collection surfaces with β variable between 0° and 90° and azimuth angles a_w variable between 0° (south) and 180° (north). Such coefficients can be held to be valid for latitudes between 35°N and 45°N with an acceptable error of less than 3 %.

For example, for an east-facing surface ($a_w = 90°$) with $\beta = 40°$, the corrective factor of annual thermal producibility is equal to 0.79. If one considers the thermal requirement and the availability of solar radiation of the previous example for the reference setting, the producibility of the considered surface becomes $0.79 \cdot 1{,}292 = 1{,}021$ kWh/year. In order to obtain the same solar fraction of the requirement, a collection surface equal to $0.62 \cdot 2{,}500/1{,}021 = 1.5$ m^2 is required.

Example 1 Figure 3 shows the roof covering of a condominium building in Rome. On each slope, the reference number, orientation, inclination and corrective factor of the annual thermal producibility, determined by means of Table 2, are reported.

Table 2 Corrective factors f_c of the annual thermal energy on the reference surface varying the inclination β and the azimuth angle a_w of the collection surface

β (°) → a_w (°)↓	0	10	20	30	40	60	90
0	0.89	0.94	0.99	1.00	0.99	0.90	0.64
22.5	0.89	0.95	0.98	0.99	0.97	0.89	0.64
45	0.89	0.93	0.95	0.95	0.93	0.84	0.62
67.5	0.89	0.91	0.91	0.90	0.87	0.78	0.57
90	0.89	0.88	0.86	0.83	0.79	0.69	0.51
112.5	0.89	0.86	0.82	0.76	0.71	0.60	0.44
135	0.89	0.84	0.77	0.70	0.63	0.50	0.36
157.5	0.89	0.83	0.74	0.65	0.57	0.42	0.30
180	0.89	0.81	0.70	0.64	0.55	0.39	0.28

Fig. 3 Roof geometry with an indication of the orientation, inclination and the annual producibility corrective factor

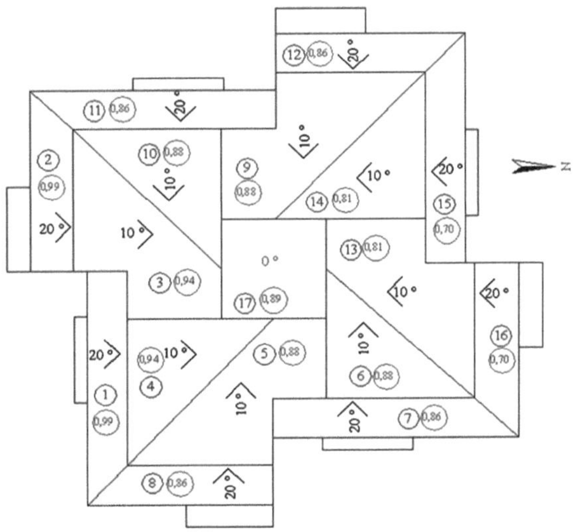

The sizing of collection surface of a centralised plant for the production of DHW for an annual thermal energy requirement of 44,200 kWh which uses evacuated heat pipe solar collectors is required. The annual incident radiation on the reference surface ($a_w = 0°$, $\beta = 30$ °C) is equal to 1,800 kWh/m^2 year. The use of modules with a collection surface equal to 2.0 m^2 is supposed.

Considering the reference setting, for a requirement of 44,200 kWh/year and an incident radiation of 1,800 kWh/m^2 year, the graphic in Fig. 2 provides a collection surface of 24 m^2 and a solar fraction of the requirement of 56 %. The annual thermal producibility results as being equal to $0.56 \cdot 44{,}200/24 = 1{,}031$ kWh/m^2.

If collectors, each with a surface area equal to 2 m^2 are used, in the hypothesis that the roof surface which can effectively be used is equal to 50 % of the effective

Table 3 Results for 17 slope calculations

Slope	N.	a_W (°)	β (°)	Area (m^2)	Effective area (m^2)	f_c	Producibility ($kWh/m^2 year$)	Installable collectors	Energy produced (kWh/year)
S	1	0	20	36.5	18.3	0.99	1021	9	18,372
	2	0	20	27.7	13.9	0.99	1021	6	12,248
	3	0	10	56.9	28.5	0.94	969	14	27,136
	4	0	10	40.5	20.3	0.94	969	10	19,383
HZ	17	–	0	39.1	19.6	0.89	918	9	16,517
E	5	90	10	52.1	26.1	0.88	907	13	23,589
	6	90	10	36.6	18.3	0.88	907	9	16,331
W	9	90	10	58.4	29.2	0.88	907	14	25,404
	10	90	10	39.1	19.6	0.88	907	9	16,331
E	7	90	20	36.9	18.5	0.86	887	9	15,960
	8	90	20	29.4	14.7	0.86	887	7	12,413
W	11	90	20	37.3	18.7	0.86	887	9	15,960
	12	90	20	29.4	14.7	0.86	887	7	12,413
N	13	180	10	56	28	0.81	835	14	23,383
	14	180	10	42	21	0.81	835	10	16,702
	15	180	20	36.6	18.3	0.70	722	9	12,991
	16	180	20	30.8	15.4	0.70	722	7	10,104

surface due to obstructions linked to the very complex roof geometry, the collection results obtainable are reported in Table 3. For each of the 17 slopes, the azimuth a_W, the inclination β, the slope surface, the usable slope surface, the corrective factor f_c of producibility, the producibility of each slope, the number of installed collectors and the energy produced are reported.

The table allows for the evaluation of all the possible solutions in order to meet the thermal requirement. For example, the use of slope n°1 and slope n°2 with a total collection surface equal to 22 m^2 (9 collectors on slope n°1 and 2 on slope n°2) is sufficient to satisfy 50 % of the thermal energy required for the production of DHW ($1{,}021 \cdot 22 = 22{,}462$ kWh). Alternatively, the production of DHW can be guaranteed using the entire available surface of slope n°3, or n°5.

4.4 A Simplified Method for the Determination of Solar Energy Utilisable for Integration in Winter Heating

The former procedure can be extended to determine the solar energy utilisable each month, integrated by the energy supplied by the heating plant, in order to meet the monthly energy requirements. Evacuated heat pipe collectors are also used in this case for the production of thermal energy at a temperature of 40–50 °C situated on the slope surfaces of the roof covering. For each month of heating, given that the monthly energy incident on the reference surface is known

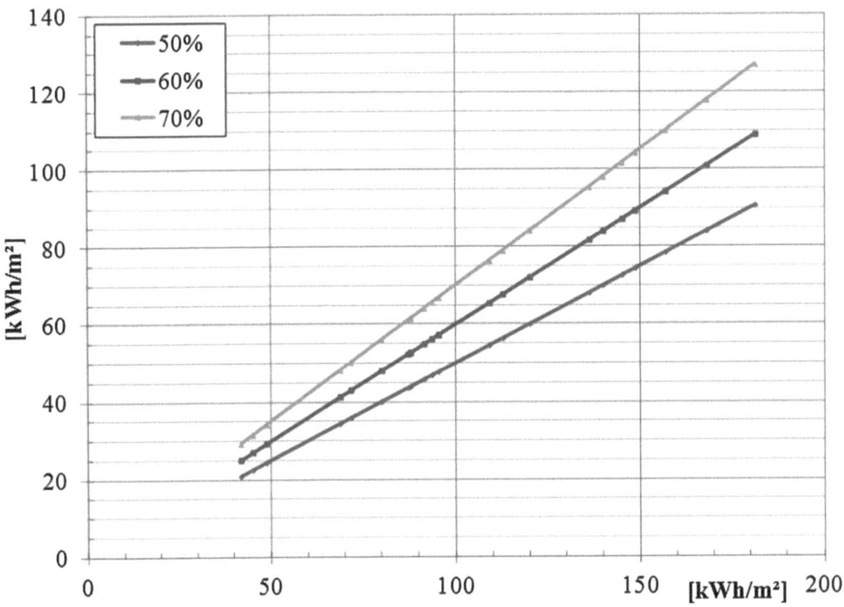

Fig. 4 Monthly energy utilisable as a function of the incident monthly energy on the reference surface ($a_w = 30°$; $\beta = 0°$) for three values of collector thermal efficiency

($a_w = 30°$; $\beta = 0°$), by means of the graphic in Fig. 4, it is possible to evaluate the monthly thermal energy available for heating (kWh/m^2), for three different values of the average monthly efficiency of the collector. For intermediate efficiency values, the energy produced can be evaluated for linear interpolation.

Given the known monthly energy utilisable on the reference plane, the utilisable energy on any tilted and oriented surface, for example on a roof covering, can be evaluated by applying monthly corrective factors reported in Table 4. Starting from the utilisable monthly energy per area unit and from the available collection areas, it is possible to determine the monthly and seasonal energy that can be used to integrate the thermal energy requirements necessary for winter heating.

5 Photovoltaic System

The vertical elements of the building shell and, even better, the covering roof surfaces can be validly used for the production of electrical energy through photovoltaic panels, to be used for the functioning of building conditioning plants and systems, with a consequent reduction in electrical energy taken from the grid.

In a photovoltaic cell, the absorbed solar power is only in part transformed into electrical power, while the rest is lost towards the outside as thermal power. The

Table 4 Corrective factors f_c of the monthly utilisable energy on the reference surface varying the inclination β and the azimuth angle a_w of the collection surface

		Inclination β (°)						
		0	10	20	30	40	60	90
October								
Azimuth a_w (°)	0	0.79	0.88	0.95	1.00	1.03	1.01	0.82
	22.5	0.79	0.87	0.94	0.98	1.00	0.98	0.78
	45	0.79	0.85	0.90	0.92	0.93	0.89	0.70
	67.5	0.79	0.82	0.84	0.84	0.83	0.77	0.59
	90	0.79	0.79	0.77	0.75	0.71	0.64	0.48
	112.5	0.79	0.75	0.70	0.64	0.59	0.50	0.37
	135	0.79	0.72	0.63	0.55	0.48	0.38	0.29
	157.5	0.79	0.69	0.58	0.48	0.39	0.31	0.25
	180	0.79	0.68	0.57	0.44	0.34	0.30	0.25
November								
Azimuth a_w (°)	0	0.72	0.83	0.93	1.00	1.05	1.08	0.94
	22.5	0.72	0.82	0.91	0.97	1.01	1.03	0.88
	45	0.72	0.79	0.85	0.89	0.91	0.89	0.73
	67.5	0.72	0.76	0.78	0.78	0.78	0.73	0.57
	90	0.72	0.71	0.69	0.66	0.64	0.56	0.43
	112.5	0.72	0.66	0.61	0.55	0.51	0.43	0.32
	135	0.72	0.63	0.53	0.46	0.40	0.33	0.26
	157.5	0.72	0.60	0.48	0.39	0.34	0.31	0.25
	180	0.72	0.59	0.46	0.36	0.34	0.31	0.25
December								
Azimuth a_w (°)	0	0.67	0.80	0.91	1.00	1.07	1.13	1.02
	22.5	0.67	0.79	0.89	0.98	1.04	1.09	0.97
	45	0.67	0.76	0.84	0.90	0.94	0.95	0.82
	67.5	0.67	0.72	0.76	0.78	0.79	0.76	0.62
	90	0.67	0.67	0.66	0.65	0.63	0.57	0.44
	112.5	0.67	0.62	0.57	0.52	0.48	0.41	0.31
	135	0.67	0.57	0.48	0.41	0.37	0.31	0.25
	157.5	0.67	0.54	0.42	0.35	0.33	0.30	0.25
	180	0.67	0.53	0.40	0.34	0.33	0.30	0.25
January								
Azimuth a_w (°)	0	0.70	0.82	0.92	1.00	1.06	1.10	0.98
	22.5	0.70	0.81	0.90	0.97	1.02	1.05	0.91
	45	0.70	0.78	0.84	0.89	0.92	0.91	0.76
	67.5	0.70	0.74	0.76	0.78	0.77	0.73	0.57
	90	0.70	0.69	0.67	0.65	0.62	0.56	0.42
	112.5	0.70	0.64	0.59	0.53	0.49	0.41	0.31
	135	0.70	0.60	0.51	0.44	0.39	0.33	0.26
	157.5	0.70	0.58	0.46	0.38	0.35	0.32	0.26
	180	0.70	0.57	0.44	0.36	0.35	0.32	0.26

(continued)

Table 4 (continued)

		Inclination β (°)						
		0	10	20	30	40	60	90
February								
Azimuth a_w (°)	0	0.75	0.86	0.94	1.00	1.04	1.05	0.88
	22.5	0.75	0.85	0.92	0.97	1.01	1.00	0.83
	45	0.75	0.82	0.87	0.91	0.92	0.89	0.72
	67.5	0.75	0.79	0.81	0.81	0.81	0.75	0.59
	90	0.75	0.75	0.73	0.71	0.68	0.61	0.46
	112.5	0.75	0.70	0.65	0.60	0.55	0.47	0.35
	135	0.75	0.67	0.58	0.50	0.44	0.36	0.28
	157.5	0.75	0.65	0.53	0.43	0.36	0.31	0.26
	180	0.75	0.64	0.51	0.39	0.34	0.31	0.26
March								
Azimuth a_w (°)	0	0.84	0.92	0.97	1.00	1.01	0.96	0.74
	22.5	0.84	0.91	0.96	0.98	0.99	0.93	0.71
	45	0.84	0.89	0.92	0.94	0.93	0.86	0.66
	67.5	0.84	0.87	0.87	0.87	0.85	0.78	0.59
	90	0.84	0.84	0.82	0.79	0.76	0.67	0.51
	112.5	0.84	0.81	0.76	0.71	0.66	0.56	0.41
	135	0.84	0.78	0.70	0.63	0.55	0.44	0.33
	157.5	0.84	0.76	0.66	0.56	0.47	0.35	0.28
	180	0.84	0.75	0.65	0.53	0.42	0.32	0.27
April								
Azimuth a_w (°)	0	0.94	0.98	1.00	1.00	0.98	0.87	0.59
	22.5	0.94	0.97	0.99	0.99	0.96	0.86	0.59
	45	0.94	0.96	0.97	0.96	0.93	0.83	0.60
	67.5	0.94	0.95	0.94	0.92	0.88	0.78	0.58
	90	0.94	0.93	0.90	0.86	0.82	0.72	0.53
	112.5	0.94	0.91	0.86	0.80	0.75	0.63	0.47
	135	0.94	0.89	0.82	0.75	0.67	0.53	0.39
	157.5	0.94	0.88	0.80	0.71	0.61	0.44	0.32
	180	0.94	0.88	0.80	0.70	0.59	0.38	0.29

equation of the instantaneous thermal balance in a stationary regime can be placed in form [2]

$$P_{cel} = A_c G_c \tau\alpha - A_c U_c (T_c - T_a) \tag{12}$$

where P_{cel} is the electrical power supplied by the cell, $\tau\alpha$ produced by the transmission coefficient of the radiation through the covering system and the absorption coefficient of the cell, U_c the coefficient of the thermal exchange between the cell and the external environment, T_c the average temperature of the cell and T_a the external air temperature.

The cell efficiency is defined as the relation between the electrical power generated and the incident solar irradiance:

$$\eta_c = \frac{P_c}{A_c G_c} \tag{13}$$

Replacing (12) with (13), the following is obtained:

$$\eta_c = \tau\alpha - \frac{U_c(T_c - T_a)}{G_c} \tag{14}$$

If one considers a module, formed by several cells which are electrically connected and closed in a sealed container, or a panel, made from more connected models and assembled in a rigid structure, for a total area A and formed by n cells, the efficiency is written in the following form:

$$\eta_c = \frac{nP_c}{G_c A} \tag{15}$$

Substituting the P_{cel} calculated with (12)

$$\eta = \frac{nA_c}{A}\eta_c = F_R\eta_c \tag{16}$$

where F_R is the fill factor, relation between the total area occupied by the cells and area A of the module or the panel.

Experimentally, it was ascertained that the efficiency of a cell (or of a module, or a panel) at temperature T_c can be expressed as [6]

$$\eta = \eta_R[1 - \beta(T_c - T_R) + \gamma \log_{10} G_c] \tag{17}$$

with η_R the efficiency of the cell evaluated in reference conditions ($G_R = 1{,}000$ W/m^2, $T_R = 298$ K, air mass $m = 1$) and G irradiation expressed in kW. For silicon cells, it can normally be assumed that $\beta \cong 0.0045\ °C^{-1}$ and $\gamma \cong 1.3$.

Often, the logarithmic term is neglected in calculations; therefore, (17) becomes

$$\eta = \eta_R[1 - \beta(T_c - T_a)] \tag{18}$$

The equation of instantaneous balance (12) can be written with reference to a panel:

$$P = AG_c\tau\alpha - AU_c(T_c - T_a) = \eta A G_c \tag{19}$$

with T_c average temperature of the panel, equal to the average temperature of the cells. Obtaining the cell temperature from (18) and substituting it in (19), the following expression is obtained for the efficiency:

$$\eta = \frac{\eta_R\left[1 - \beta(T_c - T_R) - \beta\frac{\tau\alpha G_c}{U_c}\right]}{1 - \eta_R\frac{\beta G_c}{U_c}} \tag{20}$$

which consents to the calculation of the instantaneous efficiency of the panel given the instantaneous values of the environmental temperature T_a and of the solar irradiation G_c. Often, for calculation simplicity, the denominator is assumed to be unitary. Given the known instantaneous efficiency of the cell evaluated with (20), the relation (19) permits the calculation of the instantaneous temperature of the cells.

Finally, in the energy performance calculations of a photovoltaic plant, it is useful to determine the average monthly efficiency, relation between the total energy produced and the total incident energy in the same time frame. Such an evaluation is obtained considering the average hourly values of the efficiency and irradiation according to the relation:

$$\bar{\eta} = \frac{\int_{\text{hour}} \eta G \mathrm{d}t}{\int_{\text{hour}} G \, \mathrm{d}t} = \frac{\sum \eta \, G \, \Delta t}{\sum G \, \Delta t} \tag{21}$$

with Δt time interval equal to 1 h.

The evaluation of the thermal power lost by the panel due to transmission towards the external environment requires the determination of the thermal exchange coefficient. A simple method is based on the knowledge of the nominal operative temperature of the cell Nominal Operative Cell Temperature (NOCT) [2]. This measurement represents the cell temperature, experimentally measured in its working position, in an open circuit (zero electrical power), in incident irradiation conditions equal to 800 W/m^2, wind velocity of 1 m/s and environment temperature of 20 °C. In such conditions, by applying (14), the following is obtained:

$$\frac{\tau\alpha}{U_c} = \frac{\text{NOCT} - 20}{800} \tag{22}$$

In this way, it is possible to determine the thermal exchange coefficient for the considered reference conditions, given the produced $\tau\alpha$ by the cell covering system. The NOCT value is usually provided by the constructors and assumes values which vary between 40–45 °C.

5.1 Photovoltaic Cell Types and Plant Components

One of the main limits of photovoltaic technology is linked to the high costs and to the limited efficiency of conversion of solar radiation into electrical energy. The research of materials which that ensure high electrical efficiencies and contained costs has always been developed according to different technological approaches [7]. Monocrystalline silicon ensures efficiency which varies little over time and necessitates high costs for the preparation of the monocrystalline. In order to reduce production costs, the photovoltaic industry uses cells with polycrystalline silicon as an alternative, in which the crystals are still aggregated yet with different forms and orientations. The nominal efficiency of silicon cells available on the market varies between 14 and 17.5 %.

Thin-film cells are formed by thin layers of semiconductor materials applied on a solid support. The use of a thin film notably reduces the quantity of semiconductor material necessary in the cell, compared to crystalline silicon wafers, and consequently even the production costs are reduced. The most commonly used materials in these types of cells are amorphous silicon (a-Si), gallium arsenide (GaAs), cadmium telluride (CdTe) and copper indium diselenide ($CuInSe_2$). Overall, the efficiency of thin-film cells is lower than those in crystalline silicon. Those currently available on the market present efficiency values which vary between 3.5 and 10.5 % with a lesser dependence on cell temperature due to the lower value of the temperature coefficient β which appears in the expression of the cell efficiency.

In third-generation photovoltaic cells, the most widely used materials for the creation of organic solar cells are molecular and polymeric semiconductors, such as fullerene (C60) and all its derivatives [8]. These materials are already widely used in the electronic industry, and interest lies in the simplicity and economy of the production processes. In the majority of cases, the efficiency obtained is relatively limited, currently less than 5 %, and data inherent to cell stability and energy return time are not yet available.

Given the scant power generated by a single module, it is indispensable to connect the modules in series and in parallel in order to obtain the desired current and tension values. A group of modules connected in a rigid structure is called a panel. A group of panels connected in a series in such a way as to supply nominal tension of the plant is called a string. All the panels which are connected together form the electrical energy generator, or photovoltaic field.

In order to compensate for the precariousness of the solar source, compared to the continuous requests of an electrical load, the plant can be connected in parallel to the electrical grid, or equipped with storage batteries. The maximum power tracking system "maximum power point tracker" (MPPT) allows the photovoltaic field to always operate with optimal tension and current values.

If it is necessary to have electrical energy in the form of monophase or triphase current (the modules produce continuous current), static converters, called inverters, are used. Modern MPPT are integrated with inverter devices and also carry out other functions such as the protection of loads; they realise a parallel connection between the different strings and have the function of acquiring functioning data.

5.2 Dimensioning of Photovoltaic Generators

The planning of photovoltaic plants is executed with simplified methods using different methods according to whether one is dealing with plants which are directly connected to the electrical grid (so-called grid connected, which are widespread) or plants which are for isolated users who use a battery system to store electrical energy (stand-alone). In the latter case, as well as planning the generator, it is also necessary to design the storage system to limit, as much as possible, the electrical energy produced and not consumed which needs to be dissipated in the case of battery saturation.

Grid-connected plants favourably benefit from being connected to the national grid, which can be considered as the ideal storage system to which it is to return the excess electrical energy produced.

The planning of a photovoltaic generator can be made in terms of power, or, and in the majority of cases, according to energetic criteria using the annual energy requirement as the reference parameter. The sizing of the plant requires evaluations of an economic nature, considering any eventual incentive campaigns.

There are two commonly used simplified energetic methods: the Siegel et al. method [9] and the Clark et al. method [10]. The two methods are applicable both for verification and for the basic planning of the plants. The first is valid in the case that the electrical energy produced is contemporaneously absorbed by the load, assumed to be constant. This hypothesis is true in the case of grid-connected plants, in which the eventual excess electrical power produced, compared to that absorbed by the load, is supplied to the grid. The method estimates the average monthly daily efficiency of the photovoltaic field from the average monthly daily irradiation values.

The second method considers the variable hourly profiles of the electrical load and uses the hourly irradiation values in average monthly days. In the following paragraph, a simplified model for the evaluation of the annual energy obtainable per peak kWh in the case of plants which are connected directly to the grid is shown.

5.3 Simplified Procedure for the Evaluation of Electrical Energy Producible in Buildings

In buildings which are subject to important renovation works, and also in new buildings, the evaluation of annually produced energy by photovoltaic panels can be obtained by a simplified procedure. This simplified procedure is sufficiently accurate and uses the energy values determined considering a reference placement of the panels and corrective factors to take into account their eventual different orientation and tilt [5]. The procedure lends itself to being applied in an efficient manner to buildings in the case in which the pitched roof surfaces are used for collection, which usually are differently oriented and tilted.

The calculation method uses the annual electrical power producible (kWh/kWp), determined considering a south-facing surface ($a_W = 0°$) and inclined at an angle $\beta = 30°$, with a collection area equal to that required for a nominal peak power of 1 kW. The described collection surface can be considered as a reference for latitudes between 35°N and 45°N.

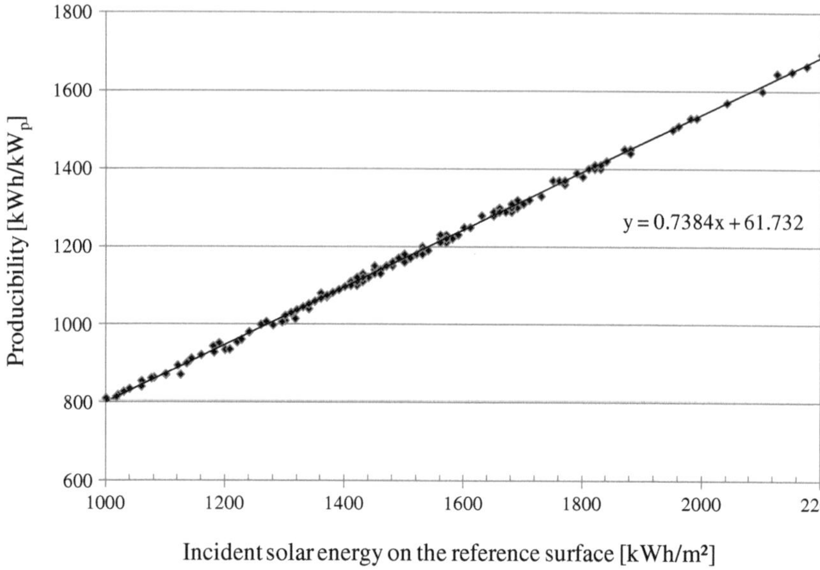

Fig. 5 Producible electrical energy on the reference plane ($a_w = 0°$, $\beta = 30°$) varying the annual solar radiation on the horizontal plane

The annual electrical energy (kWh/kWp) which can be produced is reported in Fig. 5, for mono- or polycrystalline cells, as a function of the solar energy available on the horizontal plane (kWh/m^2). The radiation data considered are those relative to the European climatic conditions. The energy which can be produced varies between 1,050 and 1,600 kWh/kWp and can be considered, with a good approximation, a linear function of the solar radiation available on the horizontal plane.

In order to evaluate the electrical producibility varying the arrangement of the collection surface, the corrective factor FC, defined as the relation between the electrical energy produced on the considered surface (a_W,β) and the corresponding energy on the reference surface, is used:

$$\mathrm{FC} = \frac{E_{\mathrm{el}}(a_W,\ \beta)}{E_{\mathrm{el}}(a_W = 0°,\ \beta = 30°)} \tag{23}$$

The FC factor values are provided by Table 5 and are valid for values of angle β which are variable between 0° and 90° and an azimuth between 0° and 180°, in such a way as to also include the photovoltaic field layout on vertical walls. The FC values can be held to be valid for latitudes between 35°N and 45°N with variances which do not exceed 3 %. Starting from the reference energy, the FC factor allows for the determination of the electrical energy produced on a surface which is oriented and inclined.

Table 5 FC corrective factor values of the reference producibility varying the grade angle β and the azimuth a_W of the collection surface

β (°) → a_W (°)↓	0	10	20	30	40	60	90
180	0.88	0.80	0.70	0.60	0.50	0.32	0.15
157.5	0.88	0.81	0.71	0.62	0.53	0.36	0.19
135	0.88	0.82	0.75	0.67	0.60	0.46	0.27
112.5	0.88	0.84	0.80	0.75	0.70	0.58	0.38
90	0.88	0.88	0.86	0.83	0.80	0.70	0.48
67.5	0.88	0.91	0.91	0.90	0.88	0.80	0.57
45	0.88	0.92	0.95	0.96	0.95	0.87	0.62
22.5	0.88	0.94	0.98	0.99	0.98	0.91	0.65
0	0.88	0.95	0.98	1.00	1.00	0.92	0.64

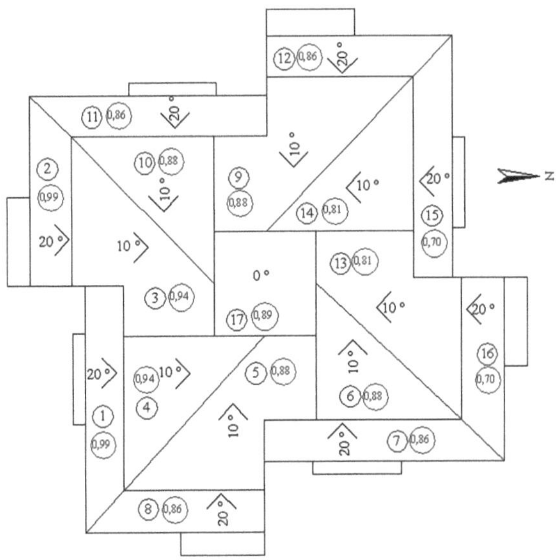

Fig. 6 Example of the roof of an existing building object of a photovoltaic installation

The procedure applied to different surfaces of the covering of a pitched roof allows for a classification of their production of electrical energy.

Example 2 Figure 6 represents a plan of the roof of building situated in Rome. The reference number, tilt angle β, azimuth a_W, and the FC corrective factor of the electrical producibility using the values from Table 5 are written on each slope. In the hypothesis that only 50 % of the slope surfaces can be used for collection, determining for a plant with peak power of 5 kW, the production of electrical energy obtainable from the different surfaces and the layout of the photovoltaic field ensures the maximum production of energy. *Data*: annual solar radiation available on horizontal plane 1,600 kWh/m^2; necessary surface area of panels for a peak power of 1 kW equal to 8 m^2.

Table 6 Calculation results

Slope	N.	a_W (°)	β (°)	Area (m^2)	Effective area (m^2)	f_c	Producibility (kWh/kW_p)	Installable power (kW)	Energy produced (kWh/year)
S	1	0	20	36.5	18	0.99	1,231	2.2	2,707
	2	0	20	27.7	14	0.99	1,231	1.7	2,092
	3	0	10	56.9	28	0.94	1,168	3.5	4,089
	4	0	10	40.5	20	0.94	1,168	2.5	2,921
HZ	17	–	0	39.1	20	0.89	1,106	2.4	2,655
E	5	90	10	52.1	26	0.88	1,094	3.2	3,500
	6	90	10	36.6	18	0.88	1,094	2.2	2,406
W	9	90	10	58.4	29	0.88	1,094	3.6	3,938
	10	90	10	39.1	20	0.88	1,094	2.4	2,625
E	7	90	20	36.9	18	0.86	1,069	2.3	2,459
	8	90	20	29.4	15	0.86	1,069	1.8	1,924
W	11	90	20	37.3	19	0.86	1,069	2.3	2,459
	12	90	20	29.4	15	0.86	1,069	1.8	1,924
N	13	180	10	56	28	0.81	1,007	3.5	3,524
	14	180	10	42	21	0.81	1,007	2.6	2,618
	15	180	20	36.6	18	0.70	870	2.2	1,914
	16	180	20	30.8	15	0.70	870	1.9	1,653

From the graphic in Fig. 5, for an availability of energy of 1,600 kWh/m^2, the reference producibility is equal to 1,243 kWh/kWp. Using the available slope surfaces, the results obtainable in the production of electrical energy are summarised in Table 6. For each slope, the following data are reported: azimuth a_w, grade β, corrective factor FC, available surface area, surface area that can be used, producibility, installable power and energy produced.

The greatest production of electrical energy is obtained using slope numbers 1, 2 and 3. For a peak power of 5 kW, it is necessary to install $5 \cdot 8 = 40$ m^2 panels.

Considering that the usable surface for each slope is as follows:

Panel surface on slope n. 1:	18 m^2
Panel surface on slope n. n. 2:	14 m^2
Panel surface on slope n. n. 3:	$40 - 18 - 14 = 8$ m^2

Therefore, the total production is $2,707 + 2,092 + \frac{8}{28} \cdot 4,089 = 5,967$ kWh/year

6 Direct Gain Passive Solar Systems

A glazed surface and the relative belonging environment form a direct gain passive system and represent the simplest and most economic system to use solar energy. In such a system, the solar radiation entering the glazed surfaces is in part absorbed by the environment walls, which have the function of thermal absorber and storage, and in part exits through the same glazed surface. The thermal level of the energy stored in the walls of a cavity depends on their dynamic properties, which determine the fraction of energy absorbed which is ceded to the indoor air. Such contributions give rise to a reduction in the winter thermal requirements and to an increase in the requirement during summer.

Another shell component which is easily realisable in buildings is the sunspace, which allows for architectural solutions which are of interest from an energy viewpoint. The control of entering radiation through the shell, by means of screens, and the possibility of using airflow rates for ventilation of the sunspace allows for the reduction in the energy requirement of the adjacent environment and, at the same time, creates acceptable thermal conditions within the sunspace.

6.1 Solar Gain Through Windows

The solar power entering in an environment through the glazed surface unit is given by the relation:

$$G_e = I_{b0} R_b \tau_b + I_{d0} R_d \tau_d + (I_{b0} + I_{d0}) R_r \tau_g \qquad (24)$$

with I_{bo} and I_{do} direct and diffuse radiation on the horizontal plane; R_b, R_d and R_r, respectively, inclination factor of direct, diffuse and reflected radiation and τ_b, τ_d and τ_g transmission coefficient of the glazed system of the direct, diffuse and reflected radiation [2].

If the instantaneous transmission coefficient of global radiation τ is introduced, the former can be expressed in the form of

$$G_e = G\tau \qquad (25)$$

with G incident solar power on the external surface of the glass and τ given by the relation:

$$\tau = \frac{I_{b0} R_b \tau_b + I_{d0} R_d \tau_d + (I_{b0} + I_{d0}) R_r \tau_g}{I_{b0} R_b + I_{d0} R_d + (I_{b0} + I_{d0}) R_r} \qquad (26)$$

The daily average monthly energy entering the glazed surface unit can be evaluated by considering the daily average monthly values of the quantities that appear in the relation (24):

$$\bar{E}_e = \bar{E} \bar{R}_b \bar{\tau}_b + \bar{D} R_d \tau_d + (\bar{B} + \bar{D}) R_r \tau_g \qquad (27)$$

Table 7 Average monthly transmission coefficient for simple clear glass (thickness = 4 mm) varying exposure

Month	Jan	Feb	Mar	Apr	May	Jun	Jul	Aug	Sep	Oct	Nov	Dec
South												
Milan	0.799	0.785	0.750	0.700	0.656	0.637	0.646	0.683	0.731	0.776	0.796	0.802
Rome	0.795	0.779	0.740	0.683	0.635	0.617	0.626	0.664	0.718	0.769	0.792	0.799
Messina	0.791	0.772	0.728	0.664	0.614	0.602	0.607	0.643	0.704	0.761	0.787	0.795
East/west												
Milan	0.706	0.731	0.747	0.753	0.754	0.753	0.753	0.753	0.749	0.736	0.713	0.697
Rome	0.714	0.734	0.748	0.753	0.753	0.751	0.752	0.752	0.750	0.739	0.720	0.706
Messina	0.720	0.737	0.748	0.752	0.752	0.750	0.751	0.752	0.750	0.740	0.724	0.713
North												
Milan	0.730	0.730	0.730	0.712	0.674	0.669	0.672	0.696	0.730	0.730	0.730	0.730
Rome	0.730	0.730	0.730	0.711	0.670	0.664	0.667	0.695	0.730	0.730	0.730	0.730
Messina	0.730	0.730	0.730	0.710	0.666	0.658	0.662	0.692	0.730	0.730	0.730	0.730

The calculation procedure of the average monthly direct radiation inclination factor $\bar{R}_b$ is reported in [2]. Alternatively, in (27), the relation can be reused:

$$\bar{E}_e = \bar{E}_i \bar{\tau} \tag{28}$$

with E_i daily average monthly energy incident on the external glazed surface and $\bar{\tau}$ transmission coefficient of the daily average monthly global radiation which can be evaluated with the relation:

$$\tau = \frac{\bar{E}\bar{R}_b\bar{\tau}_b + \bar{D}R_d\tau_d + (\bar{B}+\bar{D})R_r\tau_g}{\bar{E}\bar{R}_b + \bar{D}R_d + (\bar{B}+\bar{D})R_r} \tag{29}$$

In Tables 7 and 8, the $\bar{\tau}$ values for simple and double clear glass in clear sky conditions are reported for three Italian localities, Milan, Rome and Messina. Comparable results can be obtained for Bordeaux, Barcelona and Athens, respectively [11].

The solar radiation which is transmitted through glazed surfaces undergoes numerous reflections within the environment. Upon each reflection, the radiation attenuates due to the absorption caused by the walls, and the radiation which falls from the indoor space on the same glazed surface is, in part, dispersed externally.

The effective absorption coefficient of the environment α_{cav} is the ratio between the absorbed solar power and the entering solar power:

$$\alpha_{cav} = \frac{Q_{ass}}{A_f G_e} \tag{30}$$

The power absorbed by the space, or net solar gain, can be calculated, taking into account (31), with the relation:

$$Q_{ass} = A_f \alpha_{cav} \left[I_{b0} R_b \tau_b f_i + I_{d0} F_{r-s} \tau_d + (I_{b0} + I_{d0}) \frac{\rho}{2} \tau_g \right] \tag{31}$$

Table 8 Average monthly transmission coefficient for double clear glass (4/12/4 mm) varying exposure

Month	Jan	Feb	Mar	Apr	May	Jun	Jul	Aug	Sep	Oct	Nov	Dec
South												
Milan	0.653	0.636	0.596	0.541	0.492	0.471	0.481	0.522	0.575	0.625	0.649	0.656
Rome	0.648	0.629	0.584	0.522	0.469	0.452	0.460	0.501	0.561	0.617	0.644	0.652
Messina	0.643	0.621	0.570	0.500	0.447	0.438	0.442	0.478	0.544	0.608	0.639	0.648
East/west												
Milan	0.549	0.576	0.594	0.601	0.602	0.600	0.600	0.601	0.597	0.582	0.557	0.539
Rome	0.557	0.580	0.595	0.601	0.600	0.599	0.599	0.600	0.597	0.585	0.564	0.549
Messina	0.564	0.583	0.596	0.600	0.599	0.598	0.598	0.599	0.597	0.587	0.569	0.557
North												
Milan	0.572	0.572	0.572	0.556	0.515	0.509	0.512	0.539	0.572	0.572	0.572	0.572
Rome	0.572	0.572	0.572	0.555	0.512	0.504	0.508	0.538	0.572	0.572	0.572	0.572
Messina	0.572	0.572	0.572	0.554	0.507	0.498	0.503	0.536	0.572	0.572	0.572	0.572

- f_i is a shading factor due to the presence of external obstacles which intercept direct radiation, or of special overhangs which, during summer, reduce the solar gain. These effects are evaluated reducing the windowed surface A_f by means of the coefficient f_i;
- F_{r-s} is the view factor between the window and the sky (for a glazed surface which presents a horizontal overhang placed above it, the view factor values F_{r-s} are reported in Ref. [12]);
- F_c is a control function, which is equal to one when the glazed surface is not screened and equal to zero when an opaque screen for solar radiation is present; it is less than one when incident radiation is partially screened.

The daily average monthly solar energy absorbed by a space, due to the presence of a glazed surface, can be calculated considering the daily average monthly values of the quantities present in the relation (32):

$$\bar{E}_{ass} = A_f \bar{\alpha}_{cav}\left[\bar{B}\bar{R}_b \bar{\tau}_b \bar{f}_i + \bar{D}F_{r-s}\tau_d + (\bar{B}+\bar{D})\frac{\rho}{2}\tau_g\right]\bar{F}_c \qquad (32)$$

with $\bar{f}_i$ average monthly value of the shade factor [12] and $\bar{F}_c$ average value weighted on the radiation of the control function F_c.

6.2 Estimation of the Daily Average Monthly Absorption Coefficient of the Environments

The solar radiation which penetrates an environment through a glazed surface in part is direct radiation and in part is diffuse radiation. A simplification which is usually adopted is that of supposing solar radiation which emerges from the glazed surface, within the environment, as diffuse radiation; in this way, the directional aspects of the entering radiation are not considered. In such conditions, with reference to environments of a parallelepiped shape, with different ratios of the sides in plan and with one or more differently oriented glazed surfaces, the daily average monthly absorption coefficient of the environment α_{cav}, can be made to depend [13]

- on the average absorption coefficient of the opaque surfaces of the environment:

$$\alpha_m = \frac{\sum_i \alpha_i A_i}{\sum_i A_i} \tag{33}$$

with α_i absorption coefficient in the solar band of the surfaces $A_{i;}$
- on the glazed fraction of the environment, ratio between the glazed area and the opaque area of the cavity

$$\Psi = \frac{\sum_j A_{v,j}}{\sum_i A_i} \tag{34}$$

- on the optical properties of the glazed system, defined through the transmission coefficient of diffuse solar radiation τ_d.

The functional bond is expressed by the relation:

$$\alpha_{cav} = 1 - a\exp\left[-b\left(\frac{\alpha_m}{\Psi}\right)^2\right] \tag{35}$$

with coefficients *a, b* and *c*; quadratic functions of the transmission coefficient of diffuse radiation of the glazed system:

$$\begin{aligned} a &= 3.500 - 5.453\tau_d + 4.516\tau_d^2 \\ b &= 3.700 - 5.388\tau_d + 3.462\tau_d^2 \\ c &= 0.124 + 0.545\tau_d - 0.355\tau_d^2 \end{aligned} \tag{36}$$

The correlation (36) is valid for α_m variable between 0.20 and 0.80; Ψ inclusive between 0.025 and 0.60 and for glazed systems formed in the following manner: single clear 4-mm glass; double glazing with two 4-mm glass panes and a 16-mm air gap; and double glazing with three 2.5-mm glass panes and 12-mm air gaps. For such glazed systems, the corresponding values of the diffuse radiation transmission coefficient τ_d are, respectively, equal to 0.79, 0.59 and 0.51.

The values of the daily average monthly absorption coefficient evaluated with the previous relations, for environments with different value of the ratio between the effective absorption area and the glazed area $\alpha_m \Psi$, are reported in Table 9.

Table 9 Values of the effective absorption coefficient of environments with different glazed systems varying the parameter α_m/Ψ

	Single glass	Double glazing	Triple glazing
α_m/Ψ	α_{cav}	α_{cav}	α_{cav}
0.33	0.29	0.42	0.48
0.60	0.47	0.58	0.62
1.33	0.67	0.74	0.77
2.10	0.76	0.81	0.83
2.40	0.78	0.83	0.85
3.00	0.81	0.85	0.87
3.87	0.85	0.88	0.89
4.90	0.87	0.90	0.91
5.80	0.89	0.92	0.93
6.20	0.90	0.92	0.93
7.50	0.91	0.93	0.94
9.00	0.93	0.94	0.95
10.50	0.94	0.95	0.96
12.40	0.95	0.96	0.96
18.60	0.97	0.97	0.98
24.80	0.98	0.98	0.98

6.3 A Complete Model for the Calculation of Solar Gains in Windowed Environments

If one considers an environment with one or more glazed surfaces, the daily average monthly solar gain through the glazed surfaces, following the calculation scheme deriving from the EN 410 (2011) Standard [29], can be evaluated with the relation:

$$Q_{sol} = \sum_{k} F_{sh,k} F_{sh,sl,k} (1 - F_{F,k}) A_{w,k} G_k F_{w,k} [\tau_{b,n} \alpha_{cav} + q_i + \tau_{b,n} (1 - \alpha_{cav}) q_e]_k \Delta t \tag{37}$$

with $F_{sh,k}$ and $F_{sh,gl,k}$ which are, respectively, the reduction factors for shading produced by external elements and due to the presence of mobile screens, for the kth surface; $F_{F,k}$ factor of the area relative to the frame; $A_{w,k}$ area of the window opening; G_k solar radiation; $F_{w,k}$ average corrective factor on the radiation incidence angles; $\tau_{b,n}$ coefficient of direct solar transmission due to normal incidence; q_i "internal" secondary radiative–convective thermal exchange inwards; q_e "external" secondary radiative–convective thermal exchange outwards; and Δt time interval.

In relation (37), in the square brackets, the first term $\tau_{b,n}$ α_{cav} represents the direct optical fraction of solar radiation absorbed by the environment; the second term q_i the secondary direct fraction, produced by the absorption of the incident solar radiation from outside and the third term, the secondary indirect fraction,

Table 10 Optical and thermal parameters of the considered glazed systems

Glazed system	$g_{gl,n}$	$\tau_{b,n}$	τ_d	$\alpha_{b,n1}$	$\alpha_{b,n2}$	$\alpha_{b,n3}$	Λ_{12} (W/m^{2}K)	Λ_{23} (W/m^{2}K)	q_i	q_e
Single	0.857	0.830	0.749	0.095	–	–	–	–	0.027	0.068
Double	0.761	0.693	0.590	0.101	0.080	–	5.03	–	0.068	0.113
Triple	0.705	0.624	0.512	0.084	0.069	0.055	5.46	5.46	0.081	0.127

generated by the fraction reflected by the indoor surfaces $\tau_{b,n}$ $(1 - \alpha_{cav})$ exiting the environment through the glazed surfaces. In the case of black cavity ($\alpha_{cav} = 1$), the term linked to the secondary indirect gain is annulled.

While the effective absorption coefficient α_{cav} characterises the environment with reference to the energy entering through the glazed surface, the parameter

$$g_{n,eff} = \tau_{b,n}\alpha_{cav} + q_i + \tau_{b,n}(1 - \alpha_{cav})\, q_e \tag{38}$$

characterises the absorption of solar radiation in the system formed by the environment and by the glazed surface, with reference to the incident solar energy on the external surface of the glass. This represents the effective solar gain coefficient of the environment and is a function of the optical and thermal properties of the glazed surface and of the optic properties and the geometrical properties of the environment.

The use of the relation (38) requires the determination of parameters $\tau_{b,n}$, q_i and q_e which are obtainable from the EN ISO 13790 (2008) Standard [30]. Given the known coefficient of total solar gain for normal incidence of the glazed system, the relation

$$g_{gl,n} = \tau_{b,n} + q_i \tag{39}$$

allows the calculation of the transmission factor $\tau_{b,n}$, given that the secondary thermal exchange expressions q_i for single, double and triple glass are known. The "external" secondary thermal exchange factor q_e can be calculated, for example, for a transparent system with three panes, with the relation:

$$q_e + q_i = \alpha_{b,n1} + \alpha_{b,n2} + \alpha_{b,n3} \tag{40}$$

with $\alpha_{b,n1}$, $\alpha_{b,n2}$ and $\alpha_{b,n3}$ direct absorption factors for normal incidence angle, respectively, for the first, second and third glass of the glazed system [30].

Table 10 provides, for the glazed systems defined in the preceding paragraph, the optical parameters for normal incidence and thermal ones which intervene in the calculation: the total solar gain coefficient $g_{gl,n}$, direct $\tau_{b,n}$ and diffuse τ_d solar radiation transmission, the direct absorption factors of the glazed sheets $\alpha_{b,n}$, the thermal conductance between the glass sheets Λ, and the "internal" q_i and "external" q_e secondary thermal exchange factors.

With an increase in the number of panes that form the glazed system, the optical parameters $g_{gl,n}$, $\tau_{b,n}$ and τ_d reduce, while the internal q_i and external q_e thermal exchange factors increase.

Table 11 provides the optical and thermal properties of double glazing

$\tau_d = 0.59$	$g_{gl,n} = 0.761$	$\tau_{b,n} = 0.693$	$q_i = 0.068$	$q_e = 0.113$

Example 3 Consider an environment with a dimension, on plans, of 6 × 6 m and a height of 3 m, with three dispersant vertical walls, facing south, east and west, and the remaining opaque vertical and horizontal bordering air-conditioned environments. In the hypothesis that the external vertical walls are in part opaque, and in part glazed, in different relations, determine the effective solar gain coefficient of the environment $g_{n,eff}$ and the percentage weight of the three fractions of the solar gain. The opaque surfaces are clear with absorption coefficient in the solar band $\alpha_i = 0.30$. The optical and thermal parameters of the considered glazed systems are reported in Table 11.

For an environment with a glazed surface of 18 m^2, one obtains

$$\alpha_m = \alpha_i = 0.30$$
$$\psi = \frac{18}{126} = 0.1429$$

The coefficients of the correlations calculated with (36) are as follows:

a = 1.855	b = 1.726	c = 0.322

The absorption coefficient of the cavity evaluated with (35) is

$$\alpha_{cav} = 0.793$$

The solar gain fractions are determined as follows:

$$\tau_{b,n}\alpha_{cav} = 0.693 \cdot 0.793 = 0.549$$
$$q_i = 0.068$$
$$\tau_{b,n}(1 - \alpha_{cav})q_e = 0.693 \cdot (1 - 0.793) \cdot 0.113 = 0.016$$

Finally, the effective solar gain coefficient results as being equal to

$$g_{n,eff} = 0.549 + 0.068 + 0.016 = 0.633$$

The percentages corresponding to the three solar gain fractions are 86.7, 10.7 and 2.6 %.

The values, for the same environment and for the three considered glazed systems, which are assumed by the three gain fractions varying the glazed surface from 4.5 m^2 (one glazed surface on one wall) to 54 m^2 (three completely glazed vertical surfaces) are reported in Table 12.

Table 12 Solar gains for an environment with different fractions Ψ and for different glazed systems

Glazed system	A_f (m^2)	Ψ	α_{cav}	$\tau_{b,n}\ \alpha_{cav}$	q_i	$\tau_{b,n}\ (1-\alpha_{cav})\ q_e$	$g_{n,eff}$	$\frac{\tau_{b,n}\ \alpha_{cav}}{g_{n,eff}}$ (%)	$\frac{q_i}{g_{n,eff}}$ (%)	$\frac{\tau_{b,n}(1-\alpha_{cav})\ q_e}{g_{n,eff}}$ (%)
Single	4.5	0.0323	0.933	0.775	0.027	0.0038	0.805	96.2	3.4	0.5
	9	0.0667	0.862	0.716	0.027	0.0078	0.751	95.4	3.6	1.0
	13.5	0.1034	0.803	0.666	0.027	0.0111	0.704	94.6	3.8	1.6
	18	0.1429	0.751	0.623	0.027	0.0141	0.664	93.8	4.1	2.2
	22.5	0.1852	0.705	0.585	0.027	0.0167	0.628	93.0	4.3	2.8
	27	0.2308	0.662	0.550	0.027	0.0191	0.596	92.3	4.5	3.4
	40.5	0.3913	0.552	0.458	0.027	0.0253	0.510	89.8	5.3	5.4
	54	0.6000	0.456	0.378	0.027	0.0307	0.436	86.8	6.2	7.7
Double	4.5	0.0323	0.946	0.656	0.068	0.0042	0.728	90.1	9.3	0.6
	9	0.0667	0.887	0.615	0.068	0.0088	0.692	88.9	9.8	1.3
	13.5	0.1034	0.837	0.580	0.068	0.0128	0.661	87.8	10.3	1.9
	18	0.1429	0.793	0.549	0.068	0.0162	0.634	86.7	10.7	2.6
	22.5	0.1852	0.753	0.522	0.068	0.0193	0.609	85.7	11.2	3.2
	27	0.2308	0.717	0.497	0.068	0.0222	0.587	84.6	11.6	3.8
	40.5	0.3913	0.620	0.429	0.068	0.0298	0.527	81.5	12.9	5.6
	54	0.6000	0.534	0.370	0.068	0.0365	0.474	78.0	14.3	7.7
Triple	4.5	0.0323	0.953	0.595	0.081	0.0037	0.679	87.5	11.9	0.6
	9	0.0667	0.901	0.562	0.081	0.0079	0.651	86.3	12.4	1.2
	13.5	0.1034	0.855	0.534	0.081	0.0115	0.626	85.2	12.9	1.8
	18	0.1429	0.815	0.509	0.081	0.0146	0.604	84.2	13.4	2.4
	22.5	0.1852	0.779	0.486	0.081	0.0175	0.585	83.1	13.9	3.0
	27	0.2308	0.745	0.465	0.081	0.0202	0.566	82.1	14.3	3.6
	40.5	0.3913	0.655	0.409	0.081	0.0273	0.517	79.1	15.7	5.3
	54	0.6000	0.574	0.358	0.081	0.0337	0.473	75.7	17.1	7.1

7 Sunspaces

Sunspaces are passive solar systems which can be easily and profitably employed in the restructuring of buildings, due to the simplicity of their construction and the ease with which they can be integrated into the existing structure, for the reduction of the winter energy requirement. The energy performance in an environment adjacent to a sunspace is obtained by using the following: the direct solar gains, represented by the radiation which directly penetrates the environment through the glazed surfaces of the sunspace and of the glass dividing wall, and the indirect solar gains, produced by the fraction of solar energy which crosses the glazed shell of the sunspace and is absorbed by the opaque walls of the sunspace, with a successive transfer of energy to the air-conditioned environment, due to the presence of the sunspace in which the air temperature is generally higher than the external temperature.

The elements which determine sunspace performance are the following:

- orientation: it is preferable that the sunspace is in a south-facing position, thus ensuring greater solar gains during winter;
- the glazed system: by means of the optical and thermal properties of the glass surfaces and frame;
- the opaque elements: through the solar absorption coefficients and the thermal capacities;
- the shading systems: to reduce entering solar radiation and limit overheating of the air;
- ventilation: to remove energy from the sunspace and limit the air temperature and obtain acceptable conditions for the occupants.

During summer, sunspaces must be completely openable in order to efficiently contrast overheating of the air.

Sunspaces can be realised through different technological and formal solutions, in relation to the specific environmental, climatic and architectural contexts. It is possible to create a classification considering the sunspace elements and the wall to which it is attached, distinguishing between the following: an attached sunspace, which is developed in an external position in relation to the facade (Fig. 7); a glazed balcony, which is obtained by closing an embedded balcony (Fig. 8); and an embedded sunspace, which is in part developed inwards, and in part outwards (Fig. 9).

With regard to the extension of the solar radiation collection surface, it is possible to distinguish between entirely glazed systems and mixed glazed–opaque systems. Furthermore, the glazed system can be continuous, with a frame only on the upper and lower sides and not on the lateral sides, or with a frame which is visible on all sides.

Glazed surfaces are the most important components of sunspaces. The types of glass that are commonly used are as follows: simple clear glass, which presents optimal solar transparency qualities but poor thermal and acoustic insulation

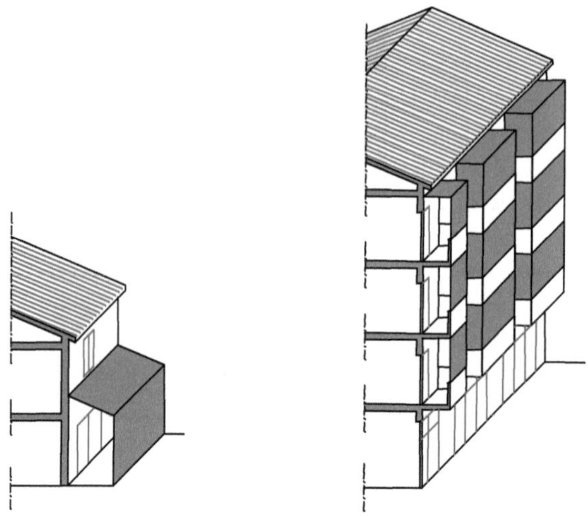

Fig. 7 Attached sunspace in a single-family building and multi-family building

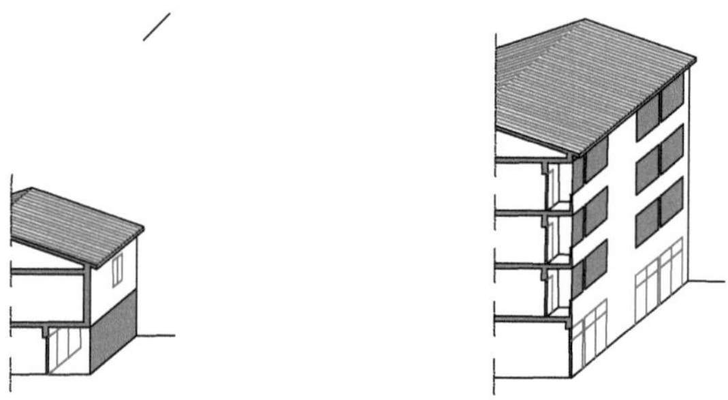

Fig. 8 Glazed balcony in a single-family and multi-family building

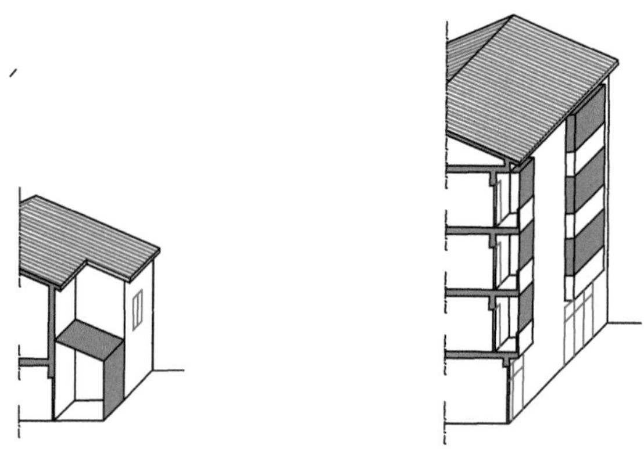

Fig. 9 Embedded sunspace in a single-family building and multi-family building

properties; double glazing, with reduced thermal transmittance values; low-emission glass, to further reduce thermal transmittance; and solar control glass, which favours the reflection of infrared solar radiation and the transmission of luminous solar radiation.

7.1 Thermal Balance of the Sunspace and Solar Gain

The solar radiation which is transmitted through the glazed shell of the sunspace is, in part, absorbed by the opaque and glass walls; in part it is lost towards the outside through the same glazed surfaces; in part it is transmitted to the adjacent environments by means of the separation elements.

The sunspace can be subject to flow rates; in such a case, it behaves as an open thermodynamic system which has exchanges of both energy and mass with the external environment and with the adjacent environments. The control of the mass and energy flows through the shell, obtained through appropriate ventilation and radiation shading strategies, allows for a reduction in the energy requirement of the adjacent building, and at the same time, creates acceptable thermal conditions within the sunspace.

The evaluation of solar gain in sunspaces requires the resolution of the optical field in the solar band, for the evaluation of the energy that is effectively absorbed by the walls, and the thermal field in the walls for the determination of the energy ceded to the internal air. For the latter evaluations, within the sunspace, it is necessary to determine the radiant field in the long infrared for the definition of the surrounding conditions.

With reference to the sunspace in Fig. 10, bordering with an adjacent environment and an underlying environment, both of which are air-conditioned, the equation of the thermal balance of the sunspace shell, formed by both opaque and glazed walls, with reference to a definite time interval, can be placed under the form:

$$Q_{as} + Q_{ai} + Q_{ae} + Q_{w} + Q_{f} = \Delta E_{w} + \Delta E_{f} \tag{41}$$

with Q_{as}energy absorbed by the sunspace walls; Q_{ai} energy that the internal surfaces exchange with the indoor air; Q_{ae} energy transferred externally; Q_{w} energy exchanged with the adjacent environment; Q_{f} energy exchanged with the underlying environment; ΔE_{w} variation of the internal energy of the wall and ΔE_{f} variation of the internal energy of the floor.

In Eq. (41), the variation of internal energy of the glazed walls was assumed to be negligible.

For the sunspace air volume, held to be a negligible thermal capacity, in the case that ventilation is realised with an external airflow rate, the balance equation is the following:

$$Q_{ai} = \dot{m}_{v} c_{p} (T_{as} - T_{ae}) \Delta t \tag{42}$$

with $\dot{m}_v$ ventilation flow rate deriving from outside; T_{as} air temperature in the sunspace; T_{ae} external air temperature and c_p specific heat of the air.

In the absence of ventilation, $\dot{m}_{v} = 0$, and the total energy Q_{ai} exchanged by the air with the internal surfaces is annulled. Such a result is obtained by means of variable exchange configurations during diurnal hours in relation to the position of the sun, while during nocturnal hours, prevalently, the sunspace air receives energy from the opaque walls and cedes energy to the glazed walls.

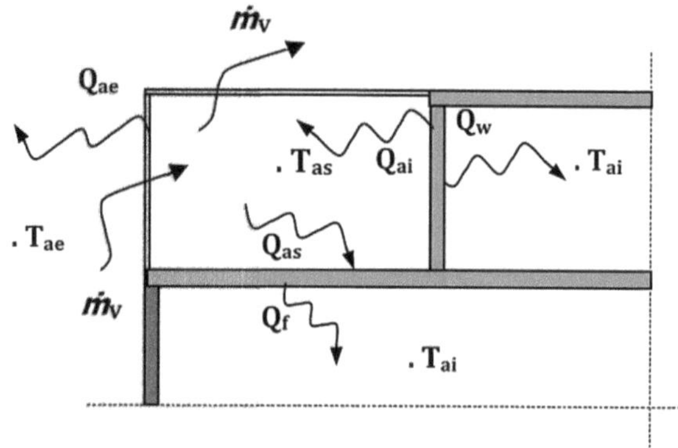

Fig. 10 Thermal exchanges between the sunspace and the adjacent environments

The energy transmitted through the sunspace shell is obtained by the radiation transmission coefficient τ_e, ratio between the transmitted energy Q_{tr} and the incident energy Q_i:

$$\tau_e = \frac{Q_{tr}}{Q_i} \tag{43}$$

The energy absorbed is evaluated with the effective absorption coefficient of the sunspace α_s, ratio between the energy absorbed $Q_{as,s}$ and that transmitted Q_{tr} [14]:

$$\alpha_s = \frac{Q_{as,s}}{Q_{tr}} \tag{44}$$

The solar gains in the sunspace can be calculated by the utilisation factor, defined by the relation

$$\eta_u = \frac{Q_{ai}^+}{Q_{as,s}} \tag{45}$$

with $Q_{as,s}$ absorbed energy and Q_{ai}^+ energy transferred by convention to the internal air, to be evaluated with the relation

$$Q_{ai}^+ = \sum S_i h_{c,i} (T_i - T_{as})^+ \Delta t \tag{46}$$

With T_i internal surface temperature of the surfaces S_i and $h_{c,i}$ convective thermal exchange coefficients. The sign + indicates that in the summation, only the positive contributions are calculated. In absence of ventilation, the energy received from the air Q_{ai}^+ is ceded prevalently to the outside. In the presence of ventilation, the net energy received for convention from the air Q_{ai} is removed by the airflow rate, as shown by the Eq. (42). In such a case, the utilisation factor imputable to ventilation $\eta_{u,v}$ is calculated as the ratio between the energy removed by the ventilation flow rate Q_{ai} and the energy absorbed $Q_{as,s}$ [15]:

$$\eta_{u,v} = \frac{Q_{ai}}{Q_{as,s}} \tag{47}$$

In order to reduce the energy requirement of the adjacent environments, the important issue is the energy transferred at a temperature of more than 20 °C, or rather the quantity

$$Q_{ai,u}^{+} = \dot{m}_v c_p (T_{as} - 20)^{+} \Delta t \tag{48}$$

in such a case the effective utilisation factor $\eta_{u,v}^{eff}$ is given by the ratio between $Q_{ai,u}^{+}$ and the absorbed energy $Q_{as,s}$

$$\eta_{u,v}^{eff} = \frac{\dot{m}_v c_p (T_{as} - 20)^{+} \Delta t}{Q_{as,s}} \tag{49}$$

Lastly, with regard to the thermal gains through the opaque walls, obtained by the adjacent environments, since the thermal flow on the wall can change directions, it is necessary to consider the relations $Q_w^{+}/Q_{as,s}$ and $Q_f^{+}/Q_{as,s}$, respectively, for the wall and the floor, with

$$Q_w^{+} = S_w h_{c,w} \left(T_{s,w} - T_{ai}\right)^{+} \Delta t \tag{50}$$

$$Q_f^{+} = S_f h_{c,f} \left(T_{s,f} - T_{ai}\right)^{+} \Delta t \tag{51}$$

with $h_{c,w}$ and $h_{c,f}$ convective coefficients of the wall and of the floor with the indoor air.

For the considered geometry, the previous relations allow for the determination of solar gains in the sunspace and in the adjacent environments, through the opaque elements and the ventilation flow rate. Such evaluations can be carried out on an hourly, daily and monthly basis.

7.2 Improvement in the Energy Performance of an Environment Adjacent to a Glazed Balcony

The optical and energy behaviour of a windowed environment is compared with that of the same environment equipped with a sunspace at the front. A total glazed sunspace was considered (Fig. 11) with different exposure (south, east/west) and situated in localities which are climatically different, Cosenza ($L = 39°18'$, Southern Italy) characterised by Mediterranean climate and Milan ($L = 45°27'$, Northern Italy) with continental climate. Similar climatic conditions, latitude, yearly solar irradiation and average monthly temperatures, are registered for Athens and Valencia in the Mediterranean area and for Belgrade and Bordeaux with continental climate.

The environment has a surface area of 24 m^2 and has a balcony in front of it, which is 6 m in length and 1.5 m in width. The closure of the balcony, by means of glazed elements, gives a sunspace with a volume equal to 27 m^3. The adjacent environment presents two dispersive external walls, with a thermal transmittance

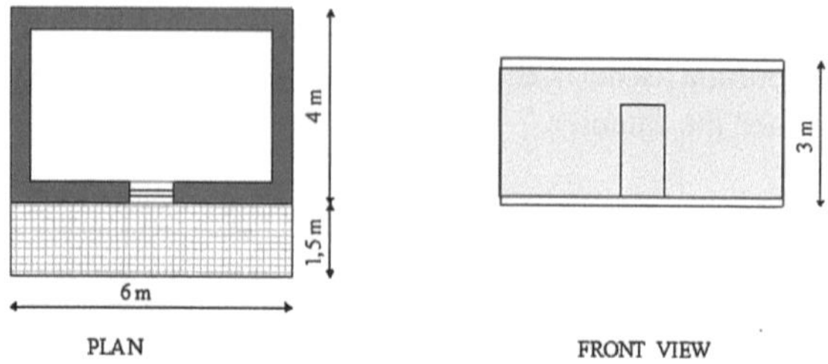

Fig. 11 Sunspace geometry—considered environment

of 0.95 W/m^2 K and a thermal capacity of 218 kJ/m^2 K, while the remaining walls border with heated environments. The combinations considered in the optical analysis are differentiated for the solar absorption coefficient of the floor α_f and of the walls α_w: $\alpha_f = \alpha_w = 0.2$ and $\alpha_f = \alpha_w = 0.5$, and furthermore $\alpha_f = 0.5$ and $\alpha_w = 0.2$ both for the sunspace and the adjacent environment. The glazed system is clear double glazing 4–12–4 mm with a thermal transmittance of $U = 2.88$ W/m^2 K and total solar transmission coefficient $g = 0.75$. The environment is separated by the environment by means of a wall with a glazed fraction which is variable between 20 and 100 %.

The presence of the sunspace gives rise to greater incident energy on the external glazed shell, to a reduction in the direct solar gain through the windowed surface of the environment and to a reduction in the thermal losses outwards, through the dividing wall, due to the increase in the air temperature in the sunspace compared to the external temperature.

The effective average monthly coefficient of the environment in absence and in presence of the sunspace, and the effective absorption coefficient of the sunspace are reported in Figs 12 and 13, varying the optical properties of the environments and of the glazed fraction of the dividing wall.

The presence of the sunspace leads to a decrease in the absorption coefficient of the environment α_{as}, due to the reduction in the absorbed energy and, to a greater extent, due to the effect of the increase in the energy entering the external glazed surface. Moreover, the presence of the sunspace introduces a significant monthly variability during winter months on decreasing the glazed fraction of the environment.

If the sunspace is facing east, in the same conditions, no significant variation of the absorption coefficients is recorded compared to southern exposure.

Figures 14 and 15 with reference to the heating period (15/11–31/3) show the energy absorbed by the environment, by the environment with the sunspace and by the sunspace, varying the glazed fraction of the dividing wall, for the three optical combinations considered.

Both for southern and eastern exposure, the reduction in energy absorbed by the environment due to the effect of the sunspace is little influenced by the glazed fraction *f* and increases with the absorption coefficient of the walls. For reflective environments, the reduction is around 30 %, and for more absorbent environments, the reduction is 35 %. With regard to the sunspace, the absorbed energy increases

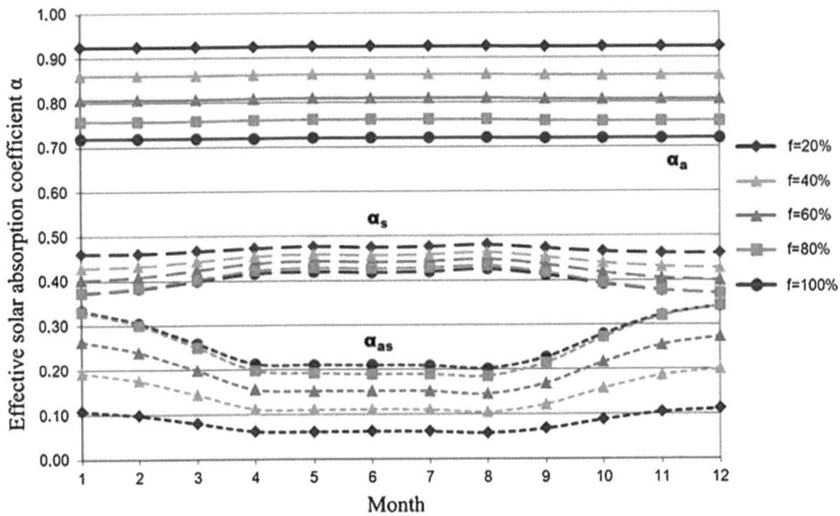

Fig. 12 Average monthly values of the absorption coefficient of the windowed environment α_a, of the environment with sunspace in front α_{as} and of the sunspace α_s, varying the glazed fraction *f*. Cosenza, southern exposure, $\alpha_f = \alpha_w = 0.2$

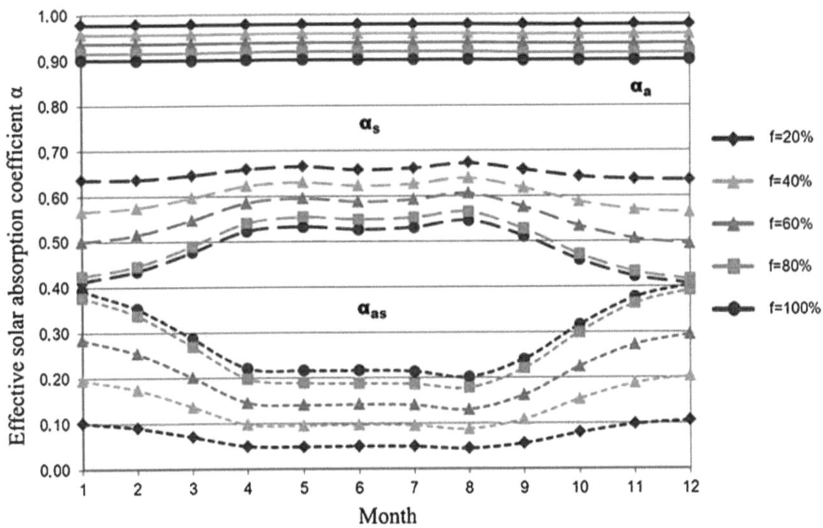

Fig. 13 Average monthly values of the absorption coefficient of the windowed environment α_a, of the environment with sunspace in front α_{as} and of the sunspace α_s, varying the glazed fraction *f*. Cosenza, southern exposure, $\alpha_f = \alpha_w = 0.5$

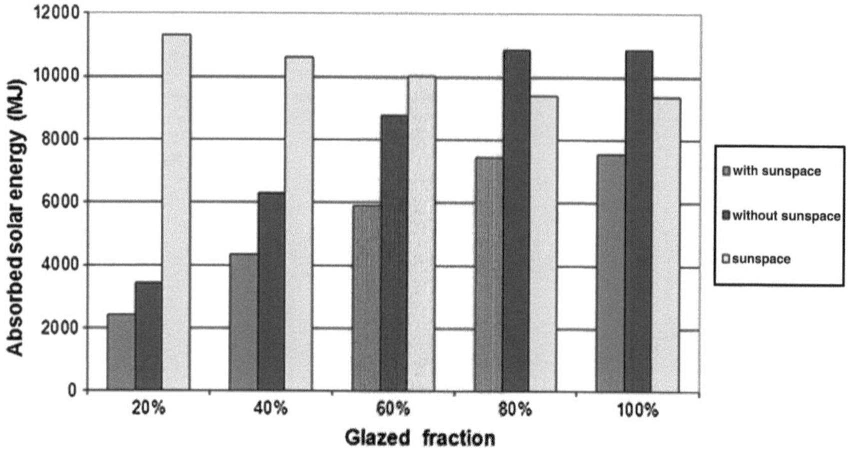

Fig. 14 Solar energy absorbed, during the heating period, by the environment, by the environment with the sunspace and by the sunspace, varying the glazed fraction. Cosenza, southern exposure, $\alpha_f = \alpha_w = 0.2$

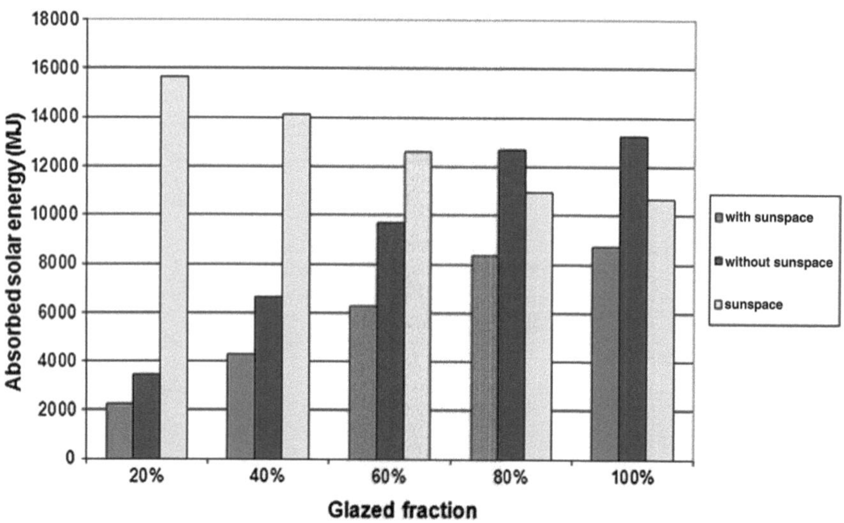

Fig. 15 Solar energy absorbed, during the heating period, by the environment, by the environment with the sunspace and by the sunspace, varying the glazed fraction. Cosenza, southern exposure, $\alpha_f = \alpha_w = 0.5$

with the absorption coefficient of the opaque surfaces and with a reduction in the glazed fraction of the dividing wall.

The prior optical analysis was repeated considering the sunspace located in Milan. The environment, in the absence of a sunspace, presents monthly absorption coefficient values which do not deviate significantly. In winter months, in the

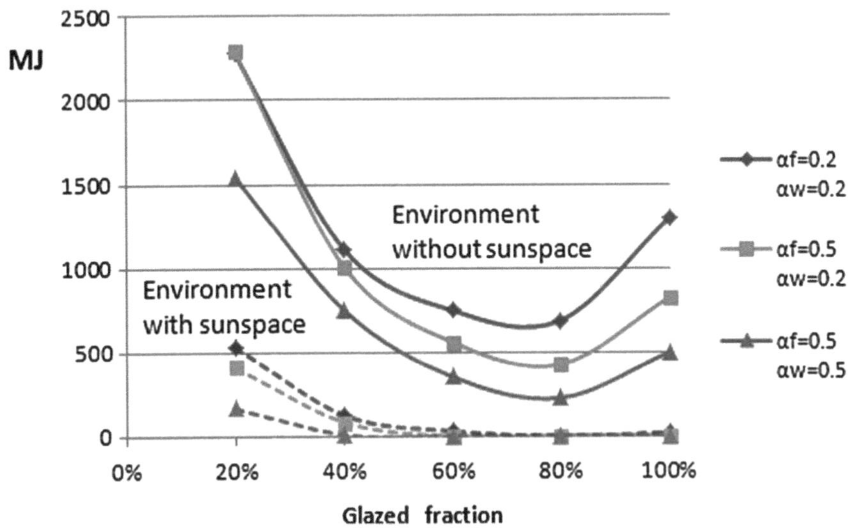

Fig. 16 Seasonal thermal requirements of the environment in the absence and presence of a sunspace, varying the glazed fraction and the optical properties of the opaque surfaces. Cosenza, southern exposure

presence of the sunspace, the environment has absorption coefficients that are about 10 % higher only in the case of environments that are moderately absorbent, while during summer months, the variations are negligible. With regard to the sunspace, the absorption coefficient can be held to be little varied with the location.

For eastern exposure, in the same conditions, the environment absorption coefficient does not present significant variations compared to the southern exposure just described, with a more contained monthly variability, as already highlighted for Cosenza.

If the energy absorbed by the environment in the presence of the sunspace is considered, the comparison, in the same optical conditions, shows a reduction in energy in Milan compared to Cosenza, which is different for higher glazed fractions. For example, for an environment with $\alpha_f = 0.5$ and $\alpha_w = 0.2$, for a glazed fraction $f = 20$ %, the reduction in absorbed energy is equal to 32 %, both for Cosenza and for Milan. With a glazed fraction $f = 100$ %, for Cosenza, there is a decrease of 34 % and for Milan a decrease of 44 %. For eastern exposure, for both localities, the reduction in absorbed energy determined by the presence of the sunspace is around 30 %, independently of the glazed fraction.

With regard to the energy requirements of the environment, in the absence of solar radiation shading systems and sunspace ventilation, with reference to the entire heating period, the results can be summarised as follows. In Fig. 16, the thermal requirements in the absence and presence of a sunspace are compared for the southern exposure.

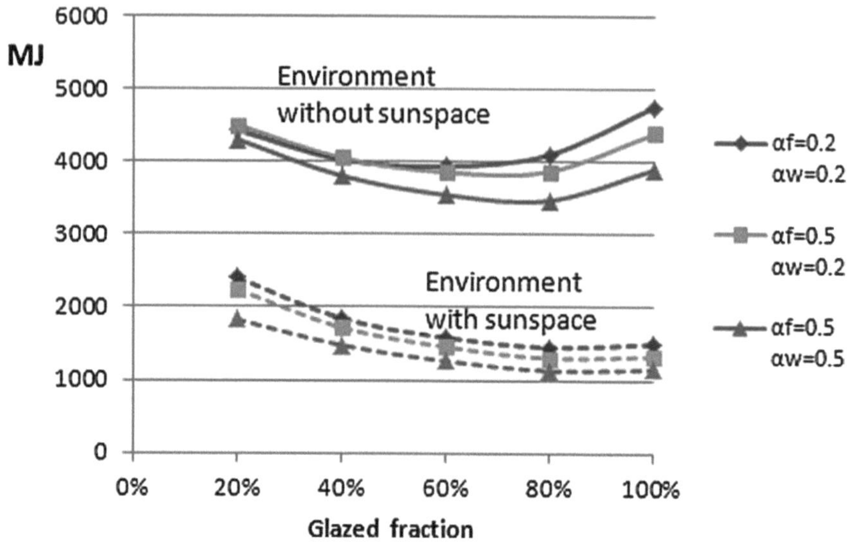

Fig. 17 Seasonal thermal requirements of the environment in the absence and presence of a sunspace, varying the glazed fraction and the optical properties of the opaque surfaces. Cosenza, eastern exposure

The presence of the sunspace requires the use of glazed fractions in the environments that do not exceed 20 % in order to avoid overheating of the environment. Such a fraction ensures a significant reduction in the winter thermal requirement of between 84 %, for reflective environments, and 89 % for environments which are more absorbent. Higher glazed fractions drastically reduce the required thermal requirement.

For eastern exposure, the trend of the environment thermal requirements is reported in Fig. 17. The reduction in the requirements is little influenced by the optical properties of the environments, and increasing the glazed fraction from 20 to 100 %, it varies between 35 and 85 %.

The figures relating to the entire period of heating (15/10–15/04) for Milan are shown in Figs 18 and 19.

The insertion of the sunspace produces important reductions in the winter thermal requirement of between 37 and 60 % for the southern exposure and between 27 and 50 % for the eastern exposure.

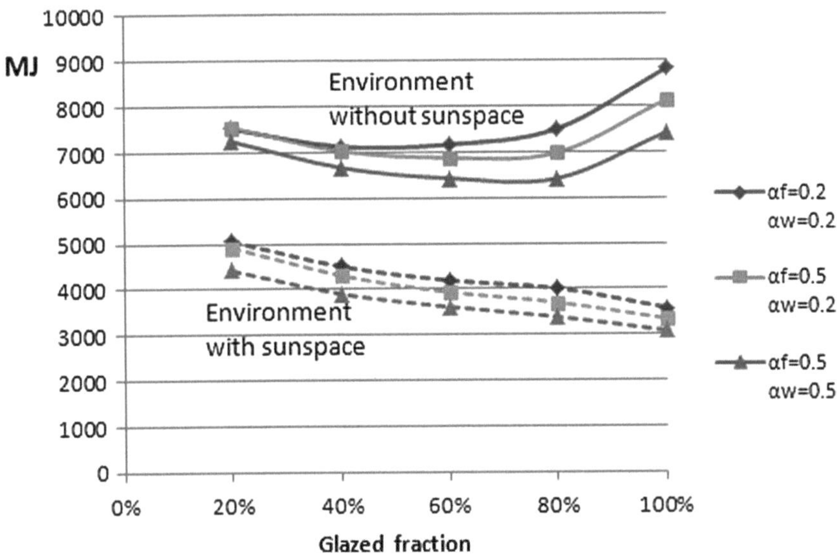

Fig. 18 Seasonal thermal requirements of the environment in the absence and presence of a sunspace, varying the glazed fraction and the optical properties of the opaque surfaces. Milan, southern exposure

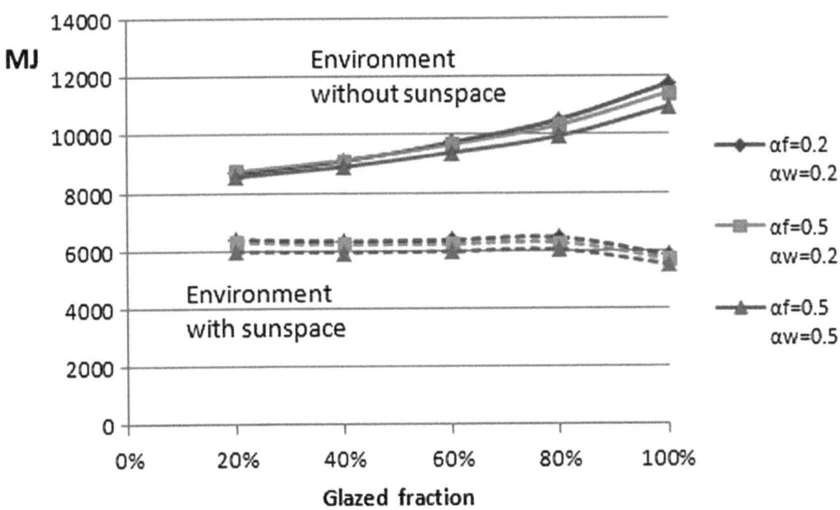

Fig. 19 Seasonal thermal requirements of the environment in the absence and presence of a sunspace, varying the glazed fraction and the optical properties of the opaque surfaces. Milan, eastern exposure

7.3 Sunspaces for Energy Requalification of a Housing Unit in an Apartment Block

The energy advantages offered by sunspaces in a requalification intervention on an existing building are evaluated. The structure is an apartment block situated in the city of Cosenza and was built in 1973. The building has seven floors, the ground floor is used for commercial premises and the six floors above are for residential use. Four housing units are present on each floor, with a surface which varies between 110 and 140 m^2. The façades are complicated by the presence of terraces, three for each apartment.

Reference was made to one housing unit, the plan of which is reported in Fig 20, situated on a middle floor with a southern and eastern exposure. The net surface on the architectural drawing is 126 m^2, with an inter-floor height of 2.70 m; it has three balconies of which two are embedded and one overhanging, all with brick parapets. In particular, the south-facing balcony has a surface area in architectural drawings of 8 m^2; those with eastern exposure have a surface area equal to 7 and 5.50 m^2.

The perimeter walls, which are hollow walls, have a thickness of 42 cm and a thermal transmittance of 0.95 W/(m^2 K); the windows, with wooden frames and single glazing, present a thermal transmittance of 4.9 W/(m^2 K); the above-lying rolling shutter boxes and the roller shutters are not insulated. The housing unit is heated with a centralised plant; DHW is produced autonomously by means of a gas boiler.

The energy requalification intervention is obtained through the conversion of the balconies into sunspaces according to two different solutions:

1. Realisation of sunspaces with transparent–opaque systems. It is the least costly and invasive intervention, which foresees the closure of the balconies by means of continuous glazing, maintaining the existing parapets;
2. Removal of the brick parapets in order to realise a system of sunspaces which are completely transparent, both in the lower part, which remains fixed, and in the upper part which can be opened.

The energetic benefits are calculated by means of the evaluation of the reduction in the winter thermal requirement obtained through thermal simulations in dynamic regime [16].

Figures 21 and 22 illustrate the two intervention hypotheses. Two solutions were considered for the realisation of the sunspaces: the use of clear single glazing 6 mm and of clear double glazing 4–12–4 mm.

In Table 13, the seasonal values of the thermal heating requirement of the housing unit in absence of and presence of the sunspace are presented. All the considered solutions give rise to a significant saving, which is variable between 31 and 41 %. The presence of double glazing offers better results, both in the absence and in the presence of the parapet.

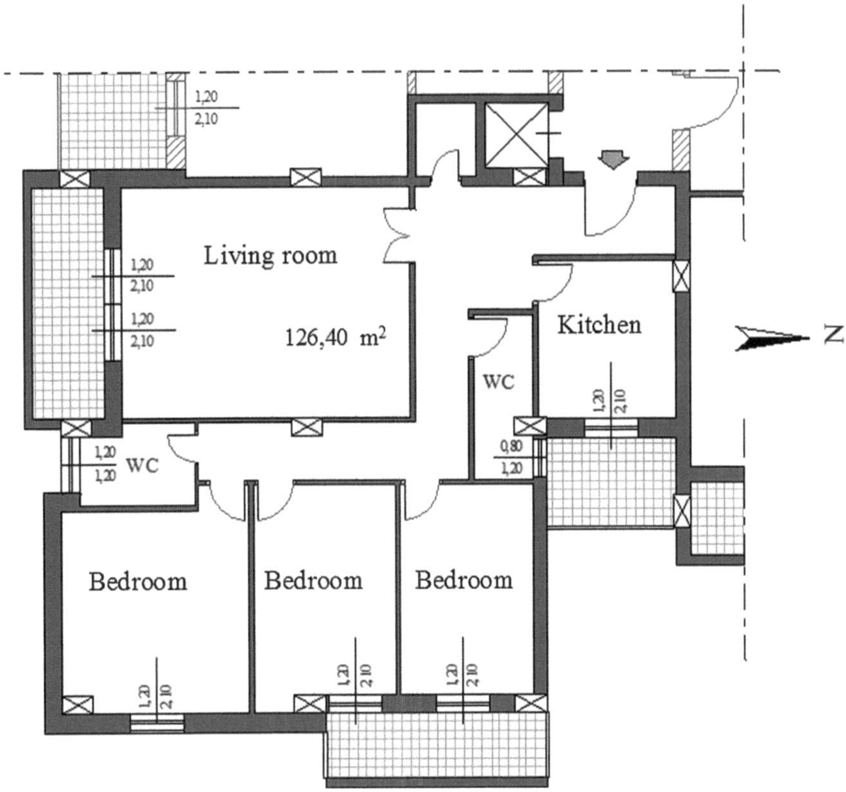

Fig. 20 Architectural drawing of dwelling with southeastern exposure

Fig. 21 Sunspaces with a mixed glass–opaque shell (solution 1)

Fig. 22 Sunspaces with an entirely glazed shell (solution 2)

Table 13 Thermal requirement of the housing unit for the heating period and reduction in percentage of the energy requirement for different intervention solutions

		Energy requirement for winter heating (kWh/m^2)	Reduction in energy requirement produced by sunspace (%)
Housing unit without sunspace		37.19	
Transparent–opaque shell	Single glazing (6 mm)	25.69	31
	Double glazing (4–12–4 mm)	23.82	36
Transparent shell	Single glazing (6 mm)	24.58	34
	Double glazing (4–12–4 mm)	21.76	41

8 Phase Change Materials

Nowadays, there is a tendency to realise buildings with light materials and small envelope thickness, so as to reduce the weight, the cost of transport and the time for construction. Unfortunately, these modern lightweight buildings suffer from pronounced overheating in summer, especially those characterised by large glazed surfaces, hence by important solar gains.

In order to compensate for the small storage capacity of lightweight buildings, the incorporation of phase change materials (PCMs) into the opaque envelope can be an effective way to enhance thermal inertia and to improve the energy performance, both in the construction stage and during refurbishment.

Indeed, thanks to their high latent heat, PCMs can store a significant amount of thermal energy at daytime while melting, thus reducing the indoor air temperature swings produced by solar and internal gains. At night, thermal energy is released and the material can restore its solid state; this stage can be enhanced by ventilating the building with fresh outdoor air.

Organic PCMs, such as paraffin, fatty acids and polyethylene glycol (PEG), are the most frequently used materials; they show good chemical stability, high latent heat and very limited super-cooling. Unfortunately, they have low thermal conductivity, which may reduce the penetration of the thermal wave into the core of the material and the full exploitation of its latent heat. Moreover, the majority of the common paraffinic PCMs are flammable, and they may not meet the strict low-flammability criteria set by the American Society for Testing Materials (ASTM). Possible solutions to limit flammability through the addition of fire retardants are discussed in Ref. [17].

The simplest and most widespread way of using PCM in buildings consists in their impregnation into gypsum, concrete or other porous materials. However, micro-encapsulation techniques have been recently developed: they consist in enclosing the PCM in microscopic polymer capsules that form a sort of powder; the powder is then included in a container made up of PVC or aluminium [18]. The final product is generally distributed as a panel or wallboard, easy to be handled and installed, from which the PCM cannot leak; furthermore, the reduced size of the microcapsules enhances the full exploitation of the PCM, because of the large surface available for heat exchange, thus optimising its effectiveness. A detailed review about the most common PCMs and the technical solutions for application in buildings can be found in [19, 20].

In order to provide a comprehensive view about the use of micro-encapsulated PCM wallboards for refurbishing lightweight buildings, so as to reduce the overheating due to solar gains, a case study is considered in the following, based on the dynamic thermal simulation of a sample building. The study will be extended to different climates in Europe and will highlight the essential role of night ventilation to maximise the effectiveness of this solution.

Moreover, with the aim of making the study more general, two different PCM wallboards will be considered.

The first wallboard (PCM-A) includes an aluminium honeycomb matrix, filled with a compound containing 60 % of a paraffin wax, encapsulated within polymeric microspheres with a diameter of approximately 5 μm. The wallboard is sealed by two thin aluminium sheets, and its overall thickness is 20 mm, as described by Ref. [21]. The weight of the wallboards is around 11 kg/m^2.

The second wallboard (PCM-B) is made of a micro-encapsulated paraffin, different from the previous one, as described later. The final form of this wallboard is a flexible panel with a thickness of 5.26 mm, covered on both sides with a very thin aluminium sheet [22]; the final weight is 4.5 kg/m^2.

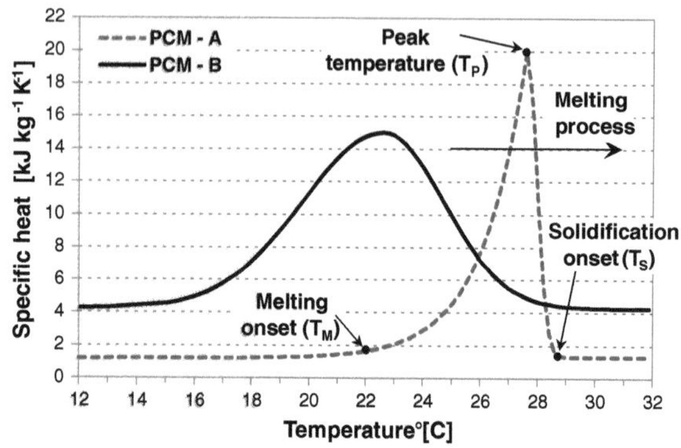

Fig. 23 Curves of the equivalent specific heat capacity for the two proposed PCMs

8.1 Characterisation of the Phase Change Materials

The melting process of a PCM used for building applications does not entirely occur at a given temperature, as for pure PCMs, but it is completed over a certain temperature range. In order to quantify the amount of heat absorbed by a PCM wallboard during phase change, the *equivalent specific heat capacity* C_{eq} is used. This parameter represents the thermal energy needed to produce a unit temperature variation of the unit mass of PCM at constant pressure:

$$C_{eq}(T) = \frac{\partial h(T)}{\partial T} \quad (52)$$

As a rule, the evaluation of C_{eq} is performed through laboratory tests, by imposing a periodic temperature fluctuation to a PCM sample and then measuring its enthalpy variation. The equivalent heat capacity normally fits a Gaussian curve, with a maximum value occurring at the peak melting temperature T_p. As an example, Fig. 23 shows the curves of the equivalent heat capacity of the two PCMs mentioned above; the corresponding mathematical formulation is reported in Ref. [23] and Ref. [24], respectively.

As one can observe, the melting process of PCM-A starts at $T_m = 22$ °C and is completed at $T_s = 28.5$ °C; the peak temperature is $T_p = 27.6$ °C, after which melting is achieved quite rapidly. Actually, according to the laboratory tests, the profile of the equivalent heat capacity during the solidification phase is slightly different, as the solidification starts at around 28 °C and the peak of the curve would occur at around 27.2 °C. Hence, the curve of the equivalent specific heat capacity for cooling/solidification is somewhat shifted towards lower temperatures if compared to that determined for the melting phase.

This behaviour, called *super-cooling*, is quite typical for paraffin, but it cannot be easily modelled by most of the programs used for the dynamic thermal simulation of buildings. However, Tabares-Velasco et al. demonstrated that the impossibility of simulating super-cooling does not affect significantly the reliability of the results [25].

On the other hand, the behaviour of PCM-B is considerably different. In fact, the melting process is distributed over a wider temperature range than for PCM-A (from $T_m = 17$ °C to $T_s = 27$ °C), but the highest value of the equivalent heat capacity is lower than for PCM-A. Furthermore, the lower peak temperature ($T_p = 22.6$ °C) suggests that the exploitation of this PCM should benefit from lower indoor temperatures.

Another important parameter that characterises a PCM is the thermal conductivity; in the case of PCM-A, the measured effective thermal conductivity is 2.7 W/(m K). This value is remarkably high, if compared to PCM-B, whose thermal conductivity varies between 0.18 and 0.22 W/(m K). This difference is imputable to the aluminium honeycomb matrix, which allows heat to be easily transferred through the panel.

On the whole, the latent heat of both PCM wallboards, i.e. the thermal energy needed to complete the whole melting process from T_m to T_s, is very similar when referred to the unit surface, as it amounts to 131.7 Wh/m^2 and to 134.0 Wh/m^2, respectively, for PCM-A and PCM-B.

8.2 Case Study

Figure 24 shows the sample building considered in this investigation: it is conceived as a module of a typical office building, with a large glazed surface protected by movable blinds, a concrete frame, a well-insulated envelope and very light partition walls to separate the different offices. This typology of buildings normally suffers from significant overheating in summer; thus, a good strategy for its refurbishment could be the application of PCM wallboards on the inner surface either of the partition walls or of the ceiling, thus enhancing the building thermal inertia.

The main façade of the building is due southwest; the size of each room is 5 m by 3.5 m, with a height of 2.5 m. Floors and ceilings are made of a non-insulated concrete slab as thick as 200 mm; the internal partitions are composed of two plasterboards with a 40-mm layer of glass wool in between. The façade has a 100-mm layer of heavyweight concrete, with an outermost layer of glass wool (70 mm). The windows are provided with an aluminium frame (10 cm in width) and a 4-16-4 double glazing with air filling. External venetian blinds are also available; these are kept open during the simulations, unless the incident solar radiation on the glazing gets higher than 250 W/m^2. The space behind the rooms at each floor is occupied by a large corridor and by a series of identical rooms facing northeast.

As far as ventilation is concerned, a constant air change rate $n = 0.5$ h^{-1} is considered for hygienic purposes. In order to check the effect of night ventilation on the performance of the PCM wallboards, an additional night ventilation rate is introduced between 21:00 and 06:00, with an air change rate $n = 4$ h^{-1} or $n = 8$ h^{-1}.

The simulations needed for this study are carried out on EnergyPlus version 7.0 over the summer period (June–September). Firstly, the weather data of Milan (Lat.

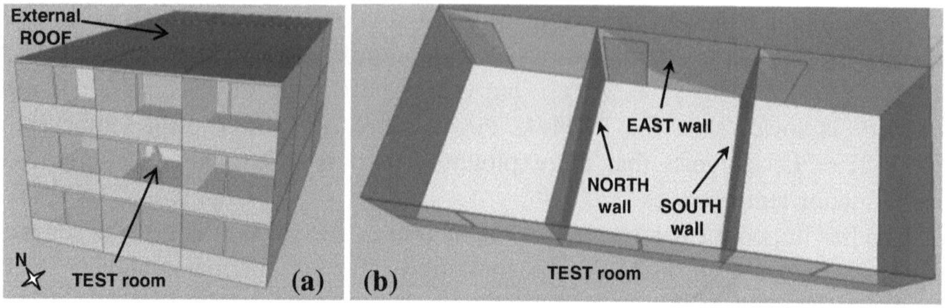

Fig. 24 **a** Model of the simulated building and **b** partitions fully covered with PCM wallboards

Table 14 Main data concerning the test room in the simulated model

Floor surface	17.5 (m^2)
Window size	1.5 × 1.7 (m^2)
Room volume	43.7 (m^3)
Partition wall: *U*-value	2.7 (W/(m^2 K))
External wall: *U*-value	0.68 (W/(m^2 K))
Floor/ceiling: *U*-value	2.8 (W/(m^2 K))
Glass *U*-value	2.7 (W/(m^2 K))
Glass *g*-value	0.76 (-)
Occupancy rate	0.12 (people/m^2)
Occupancy time	08:00–18:00
People sensible thermal load	60 (W/person)
Electric appliances	100 (W)
Lights	8 (W/m^2)

45°27'N, Italy) will be used, as an example of continental climate; in a second stage, other locations in Europe will be considered.

Three series of simulations are performed: the first one without PCM wallboards and the others with the PCM wallboards placed in the test room, respectively:

1. on the inner faces of the three partition walls,
2. on the inner surface of the ceiling, as shown in Fig. 24.

The main data used to build the simulated model are reported in Table 14.

8.3 Results and Discussion

In most of the research works available in the scientific literature and regarding the use of PCM wallboards for improving summer thermal comfort in buildings, the effectiveness of PCMs is measured by the indoor temperature drop achieved, during a short representative period, thanks to the application of the PCM, in comparison with the case without PCM. However, it is to remark that the room

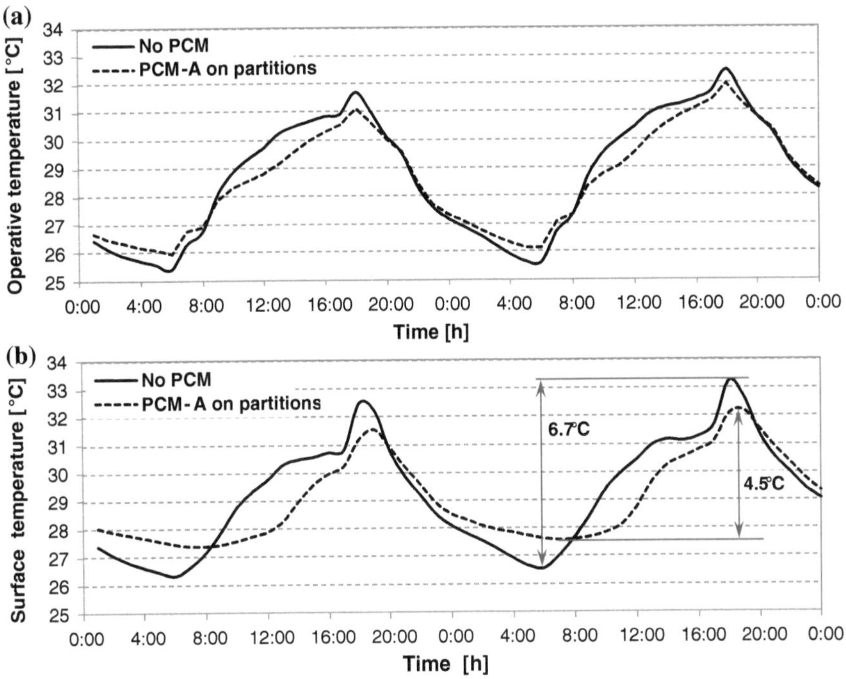

Fig. 25 Effect of PCM-A on the operative temperature (**a**) and the surface temperature of the south wall (**b**), when the night air change rate is $n = 4\ h^{-1}$

operative temperature, and not the indoor air temperature, is the parameter that directly affects the comfort sensation, as suggested by well-established comfort theories, reported in the International Standards ISO 7730 [31] and EN 15251 [32]. According to this approach, the discussion of the results will be based on the values of the operative temperature obtained from the simulations.

8.3.1 Results

The profile of the indoor operative temperature in the test room with and without PCM for two sunny days in July is shown in Fig. 25a. Both curves refer to the case with $n = 4\ h^{-1}$; in the case with PCM, the wallboards containing PCM-A are applied on the inner surface of all the partition walls.

The figure shows that the installation of the PCM wallboards yields a reduction in the peak operative temperature of about 0.7 °C, but in the central hours of the day, this difference keeps around 1.0 °C for almost 3 h. In addition, Fig. 25b shows a significant attenuation in the daily surface temperature swing of the partitions when using PCM: as an example, the peak surface temperature of the south wall drops from 33.2 to 32.1 °C during the second day, and the daily

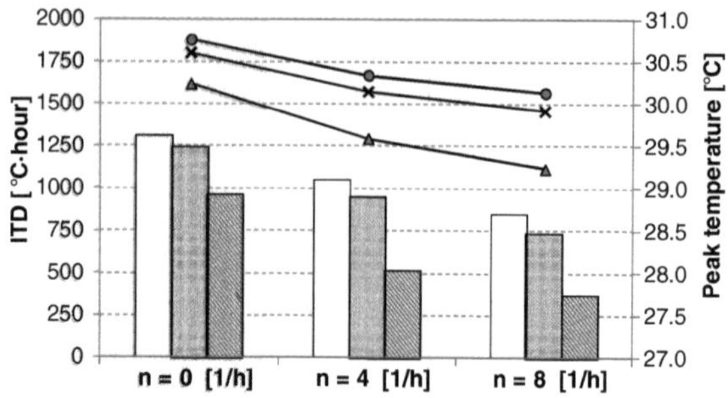

Fig. 26 Comparison based on the ITD and the average peak operative temperature, measured over the whole summer season in Milan

temperature swing is reduced from 6.7 to 4.5 °C. Finally, a slight time shift of the peak surface temperature can be observed, which is another visible effect of the additional thermal inertia provided by the PCM.

Methodological remarks

However, in the attempt of having a more comprehensive view about the effect of PCMs on thermal comfort, a deeper analysis is requested. To this aim, an effective way to quantify the uncomfortable thermal sensation due to overheating might be the measure of the difference between the operative temperature and a threshold value for it. On this basis, the indicator called *Intensity of Thermal Discomfort* (ITD) can be adopted: it is defined as the time integral, over the occupancy period P, of the positive difference between the current operative temperature and the upper threshold for comfort [23]:

$$\mathrm{ITD} = \int_{P} \left(T_{\mathrm{op}}(\tau) - T_{\mathrm{lim}}\right)^{+} \mathrm{d}\tau \tag{53}$$

Thus, the ITD measures at the same time the intensity and the duration of the thermal discomfort perceived by the occupants. The value of the threshold temperature T_{lim} depends on the choice of a specific thermal comfort theory. In this work, the adaptive approach is used, as described in the EN 15251 Standard [32]; hence, the threshold value is not constant in time, but it is determined daily as a function of the running mean outdoor air temperature [26].

Further results

At this point, the new indicators introduced so far can be used to provide some more information about the effectiveness of the wallboards containing PCM-A. In particular, the aim is to underline the role of the night ventilation as well as the effect of the position of the wallboards. The results of this analysis are reported in Fig. 26.

The first message conveyed by Fig. 26 is that the PCM wallboards are far more efficient if applied on the partition walls than on the ceiling. As an example, with reference to $n = 4\ \mathrm{h}^{-1}$, the application of PCM-A on the partition surface would yield a reduction of 51.5 % in the seasonal ITD if compared to the case without PCM, whereas such improvement would only amount to 8.9 % in the case of the ceiling. The difference between the two cases is remarkable.

Moreover, in terms of average peak operative temperature, the attenuation introduced by the PCM wallboards, still under the hypothesis $n = 4\ h^{-1}$, would be 0.7 and 0.2 °C, respectively, with the PCM applied either on the inner walls or on the ceiling. These results can be partially justified by the larger surface area available on the partitions (27 m^2) than on the ceiling (17.5 m^2). However, it is also to remark that the partitions—but not the ceiling—are directly hit by the solar radiation during the day; this determines a faster and more intense melting of the PCM, hence a more effective exploitation of its latent heat. One more simulation was devoted to the case where all the surfaces are covered by PCM (partitions plus ceiling), and the results were very close to those obtained with the PCM only on the partitions walls.

Another key message emerging from the simulations is the importance of a good night ventilation strategy that boosts the discharge of the heat absorbed during daytime. Actually, Fig. 26 shows that the peak temperature is relatively sensitive to the nigh ventilation rate, but even more sensitive to n is the ITD.

However, it can be observed that a saturation effect occurs; thus, it is not recommendable to go beyond $n = 8\ h^{-1}$, since the benefit on both the peak operative temperature and the ITD tends to vanish. It is also understood that such intense ventilation can only be practised in tertiary buildings that are supposed to be not occupied at night; furthermore, if the ventilation is procured by mechanical means, an accurate calculation of the electricity consumption should be done to avoid severe penalties to the benefits discussed so far.

8.3.2 Correlation Between Climate and PCM Effectiveness

The results reported in the previous section suggest that the wallboards containing PCM-A can allow a significant improvement in the indoor thermal comfort in a room of a typical lightweight office building located in Milan (Italy, continental climate).

Now, it is interesting to investigate whether such favourable outcomes hold true also in other climatic contexts; furthermore, the effectiveness of PCM-B in place of PCM-A has to be examined. The main results of this analysis are reported in Fig. 27 and will be discussed hereafter.

Effect of the climatic conditions

As shown in Fig. 27, the effectiveness of PCM-A in reducing the ITD in the test room is not the same for all the sites considered. Apparently, a certain correlation emerges between the ITD reduction and the latitude of the site. Indeed, the highest values of the ITD reduction occur in northern Europe (66.2 %, Paris), whereas the least satisfying results are those concerning southern Europe: around 42 % in Catania (Southern Italy) and Madrid (Spain). In addition, an average effectiveness of the PCM is observed in central Europe (51.5 %, Milan), as already remarked in Fig. (26).

Such a tendency can be easily justified if one thinks that the high values of the solar irradiance occurring in summer at low latitudes induce an intense PCM melting at daytime. Then, at night, as the outdoor air temperature is on average

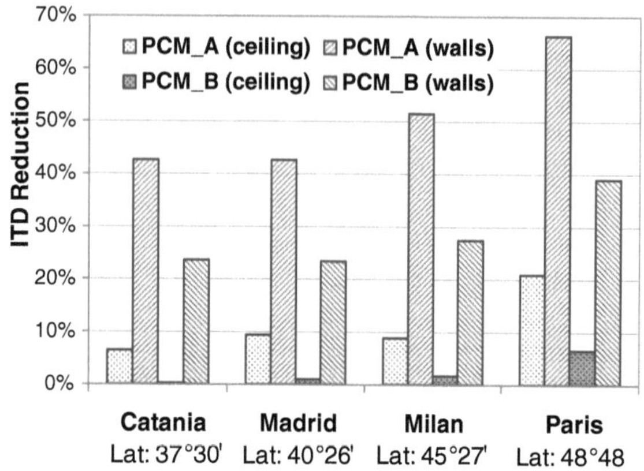

Fig. 27 Reduction in the ITD, thanks to the application of PCM wallboards on the room surfaces (all cases refer to $n = 4\ h^{-1}$)

quite high, there is a low potential for refreshing the room and discharging the heat absorbed by the PCM, which cannot be completely solidified. On the contrary, at high latitudes, the PCM melting process might be often not complete, whereas solidification at night is easily achieved. In any case, the PCM installed on the ceiling is far less effective than when applied on the partitions, whatever the latitude of the site.

Comparison between different PCMs

The results reported in Fig. 27 also show that the wallboards containing PCM-B are not as effective as PCM-A in reducing the intensity and the duration of the thermal discomfort perceived in the test room.

As an example, in the case of the PCM wallboards applied on the partitions, the reduction in the ITD remarked in Paris in comparison with the case without PCM is 66.3 % for PCM-A and only 39.1 % for PCM-B. Similar trends emerge in the other sites, as well as when the PCM is applied on the ceiling.

Here, one can observe that the phase change for PCM-B occurs over a range of temperatures quite lower than for PCM-A (see Fig. 23). Hence, in the presence of the high temperatures usually measured in summer in freerunning lightweight buildings, the storage capacity of PCM-B cannot be exploited as effectively as for PCM-A. Further investigations on this point are presented in the following.

8.3.3 The PCM Storage Efficiency

In order to assess more accurately the performance of a PCM, it is necessary to understand whether and to what extent its latent heat is effectively exploited. To this aim, it can be useful to calculate the *Frequency of Activation* (FA), i.e. the percentage of time within a given period during which the PCM is actually undergoing phase change [23]. This occurs between $T_m = 22$ °C and $T_s = 28.5$ °C for PCM-A, and in the range 17 °C – 27 °C for PCM-B (see Fig. 23).

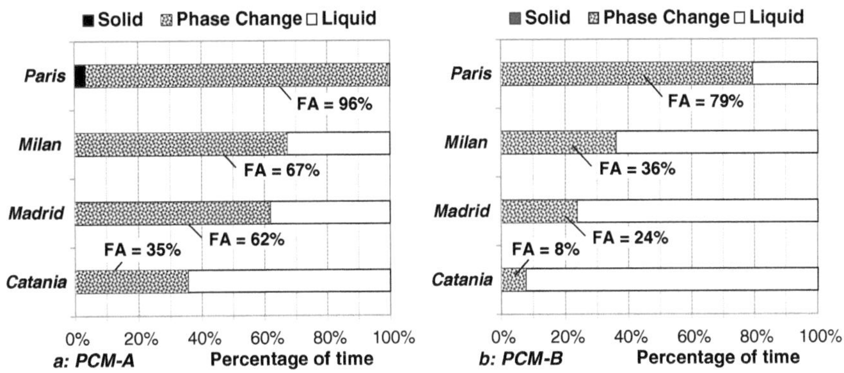

Fig. 28 Average FA of the PCM wallboards (July and August, $n = 4\ \mathrm{h}^{-1}$)

The FA can provide important information: when its value is low, it means that the PCM remains in its liquid or solid phase for a long time; thus, its latent heat capacity is not exploited. An ideal application of a PCM should imply FA = 100 %, but this is not easy to accomplish, as the activation of the PCM is highly influenced by many circumstances.

The results obtained for the sample building in terms of FA are shown in Fig. 28. Here, it is possible to observe that the wallboards containing PCM-A seem to work very well in Paris, as the PCM is almost always activated throughout the season (FA = 96 %). On the contrary, the same wallboards keep very often in the liquid phase if installed in Catania, where the FA is much lower than in other sites (FA = 35 %), due to the severe climatic conditions in summer. As concerns PCM-B, its FA is always far lower than for PCM-A.

However, not all the conditions inside the melting range have the same importance from an energetic point of view. In other words, the heat capacity of a PCM is strongly dependent on temperature: as an example, the equivalent specific heat capacity of PCM-A at the peak temperature $T_p = 27.6$ °C is almost 5 times as high as at 25 °C, and vice versa. Consequently, at 25 °C, the PCM-A, despite being activated, has a storing capacity 5 times lower than at $T_p = 27.6$ °C. Nevertheless, the FA itself is obviously not capable of accounting for this difference.

Therefore, it appears suitable to introduce a new indicator called PCM storage efficiency that measures the ratio of the thermal energy actually stored by the PCM to its maximum storage capacity, i.e. its latent heat L, as defined in Eq. (54). Since the PCM is subject to daily temperature cycles, the actual energy storage must be evaluated over the period $P = 24$ h [23].

$$\eta_{\mathrm{PCM}} = \frac{E_{\mathrm{st}}}{L} = \frac{\int_P \left(M \cdot C_{\mathrm{eq}} \cdot \frac{\mathrm{d}T_{\mathrm{PCM}}}{\mathrm{d}\tau}\right)\mathrm{d}\tau}{\int_{T_{\mathrm{m}}}^{T_{\mathrm{s}}} M \cdot C_{\mathrm{eq}}(T)\mathrm{d}T} \quad (54)$$

Table 15 Average values of the PCM storage efficiency (July and August, $n = 4\ h^{-1}$)

	PCM type A		PCM type B	
	On walls	On ceiling	On walls	On ceiling
Catania	35.0 %	9.9 %	19.7 %	7.6 %
Madrid	41.0 %	14.0 %	19.9 %	7.6 %
Milan	42.1 %	14.6 %	24.4 %	9.4 %
Paris	42.6 %	20.7 %	35.1 %	13.9 %

The average values of the daily PCM storage efficiency η_{PCM} are reported in Table 15. Here, a distinction is also made between the wallboards applied on the partitions and those installed on the ceiling. The values of this indicator are usually significantly lower than FA: as an example, the average storage efficiency is $\eta_{PCM} = 42.6\ \%$ for PCM-A in Paris, even if the PCM turns out to be activated for more than 90 % of time (FA = 96 %). Very low values of η_{PCM} occur for PCM-B, ranging from 19.7 to 35.1 % when installed on walls and from 7.6 to 13.9 % when installed on the ceiling. According to these figures, the effectiveness of the PCM wallboards seems not to be very satisfying, despite their frequent activation. The reason for this apparently reduced PCM potential can be found by looking at Fig. 29, where each point describes the mean operating conditions of a wallboard throughout a day in summer.

The highest values of the daily storage efficiency of PCM-A (between 60 and 70 %, see Fig. 29a) pertain to those days where the average temperature of the PCM wallboards is very close to the peak melting temperature. This corresponds to what already remarked by Neeper [27]. However, such a condition occurs only occasionally in July and August; on the contrary, the daily temperature of the PCM is frequently either too low (in Paris) or too high (in Catania). As a general rule, the farther from T_p is the daily average PCM temperature, the lower is the daily PCM storage efficiency.

When looking at PCM-B, the situation is more unsatisfying: here, all points regarding Catania are very close to the upper limit of the melting range; thus, the daily storage efficiency always keeps between 15 and 25 % (see Fig. 29b). Better results are observed in Paris; nevertheless, the daily values of η_{PCM} hardly exceed 50 %.

The indications coming out from this analysis are coherent with the results reported in Fig. 27. This confirms that the ability of a PCM wallboard to improve summer thermal comfort in lightweight buildings is strictly related to the possibility of exploiting its latent heat capacity; this implies the need of keeping its temperature very close to the peak melting temperature T_p as long as possible. To this aim, different techniques are currently being investigated, mostly based on the improvement of the heat transfer coefficient between indoor air and PCM wallboard.

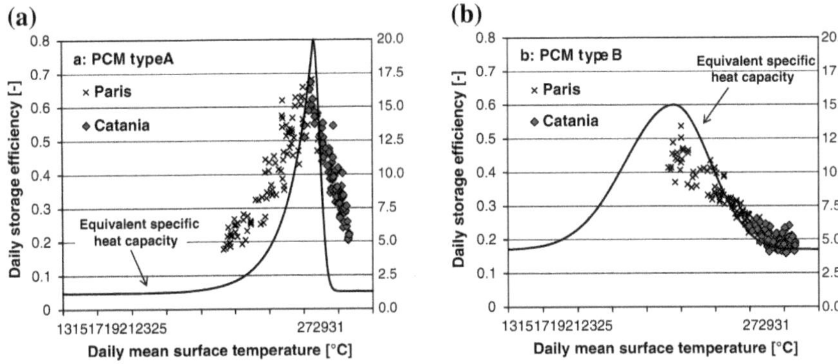

Fig. 29 Average daily PCM temperature versus average daily storage efficiency. Each point represents the mean value for all partitions (July and August, $n = 4\ h^{-1}$)

As a conclusion, the convenience of PCMs in buildings must be carefully considered, also in relation to the local climate. The main parameters that affect the PCM effectiveness are as follows:

1. the position of the PCM wallboard within the room,
2. the rate of ventilation at night, and
3. the value of the peak melting temperature for the specific PCM.

References

1. Kreith F, Kreider JF (1978) Principles of solar engineering. McGraw-Hill, New York
2. Duffie JA, Beckman WA (2006) Solar engineering of thermal processes, 3rd edn. Wiley, New York
3. Cucumo MA, Marinelli V, Oliveti G (1994) Ingegneria solare principi and applicazioni (principles and applications of solar engineering). Pitagora, Bologna
4. Eicher U (2003) Solar technologies for buildings. Wiley, England
5. Oliveti G, Arcuri N, Bruno R, De Simone M (2011) Sistemi solari nella riqualificazione energetica degli edifici (solar systems for the buildings energy requalification). National Journal La Termotecnica 5
6. Evans DL (1981) Simplified method for predicting photovoltaic array output. Sol Energy 27:555
7. Biagini MV, Lorenzini G, Saro O (2012) The evolution of photovoltaic technologies for energy production: state–of–the–art context and recent developments. Int J Energ Technol 4(6):1–11
8. Coiante D (2009) Le Nuovi Fonti di Energia Rinnovabile (The new sources of renewable energy). Franco Angeli, Milano
9. Siegel MD, Klein SA, Beckman WA (1981) A simplified method for estimating the monthly-average performance of photovoltaic systems. Sol Energy 26:413
10. Clark DR, Klein SA, Beckman WA (1984) A method for estimating the performance of photovoltaic systems. Sol Energy 33:551
11. Cucumo MA, Kaliakatsos D, Marinelli V (1993) Sul Calcolo del Coefficiente di Trasmissione Solare delle Superfici Vetrate (On the calculation of the solar transmission coefficient of glazed surfaces). In: Proceedings of the 48° National Conference A.T.I

12. Utzinger MD, Klein SA (1979) A method of estimating monthly average solar radiation on shaded receivers. Sol Energy 23:369
13. Oliveti G, Arcuri N, Bruno R, De Simone M (2011) An accurate calculation model of solar heat gain through glazed surfaces. Energy Build 43:269–274
14. Oliveti G, De Simone M, Ruffolo S (2008) Evaluation of the absorption coefficient for solar radiation in sunspaces and windowed rooms. Sol Energy 82:212–219
15. Oliveti G, Arcuri N, De Simone M, Bruno R (2012) Solar heat gains and operative temperature in attached sunspaces. Renew Energy 39:241–249
16. DEROB-LTH v1.0. (2004) User's manual. Division of Energy and Building Design, Lund Institute of Technology, Sweden
17. Sittisart P, Farid MM (2011) Fire retardants for phase change materials. Appl Energy 88:3140–3145
18. Tyagi VV, Kaushik SC, Tyagi SK, Akiyama T (2011) Development of phase change materials based microencapsulated technology for buildings: a review. Renew Sust Energ Rev 15:1373–1391
19. Rodriguez-Ubinas E, Ruiz-Valero L, Vega S, Neila J (2012) Applications of phase change materials in highly energy-efficient houses. Energy Build 50:49–62
20. Soares N, Costa JJ, Gaspar AR, Santos P (2013) Review of passive PCM latent heat thermal energy storage systems towards buildings' energy efficiency. Energy Build 59:82–103
21. Hasse C, Grenet M, Bontemps A, Dendievel R, Sallée H (2011) Realization, test and modelling of honeycomb wallboards containing a phase change material. Energy Build 43:232–238
22. Kuznik F, Virgone J, Roux JJ (2008) Energetic efficiency of room wall containing PCM wallboard: a full-scale experimental investigation. Energy Build 40:148–156
23. Evola G, Marletta L, Sicurella F (2013) A methodology for investigating the effectiveness of PCM wallboards for summer thermal comfort in buildings. Build Environ 59:517–527
24. Kuznik F, Virgone J (2009) Experimental assessment of a phase change material for wall building use. Appl Energy 86:2038–2046
25. Tabares-Velasco PC, Christensen C, Bianchi M (2012) Verification and validation of EnergyPlus phase change material model for opaque wall assemblies. Build Environ 54:186–196
26. Sicurella F, Evola G, Wurtz E (2012) A statistical approach for the evaluation of thermal and visual comfort in free-running buildings. Energy Build 47:402–410
27. Neeper DA (2000) Thermal dynamics of wallboard with latent heat storage. Sol Energy 68:393–403

Reference Standards

28. EN 15316-4-3 (2007) Heating systems in buildings—Method for calculation of system energy requirements and system efficiencies—Part 4–3: Heat generation systems, thermal solar systems
29. EN 410 (2011) Glass in building—Determination of luminous and solar characteristics of glazing
30. EN ISO 13790 (2008) Energy performance of buildings—Calculation of energy use for space heating and cooling
31. ISO 7730 (2005) Ergonomics of the thermal environment—Analytical determination and interpretation of thermal comfort using calculation of the PMV and PPD indices and local thermal comfort criteria
32. EN 15251 (2007) Indoor environmental input parameters for design and assessment of energy performance of buildings addressing indoor air quality, thermal environment, lighting and acoustics

The Calibration Process of Building Energy Models

Roberta Pernetti, Alessandro Prada and Paolo Baggio

Abstract The importance of model calibration has been growing up as a result of the energy refurbishment policy promoted by the recast Energy Performance of Buildings Directive (EPBD 2010/31/EU). In fact, with the purpose of ensuring a suitable refurbishment design with effective energy conservation measures (ECM), an accurate model has to be defined in order to assess the energy behaviour of the as-built building. In this chapter, some issues related to the model calibration are presented, starting from the definition of an operative procedure step by step. Furthermore, for the most critical phases of the procedure, analysis techniques and experimental methods are described both through theory and practical examples. Finally, throughout the chapter, the analysis of a case study is presented.

Nomenclature

CDH_{26}	Cooling degree hours at a base temperature of 26 °C
F	Sensitivity index for factorial method
HDH_{18}	Heating degree hours at a base temperature of 18 °C
k	Specific heat capacitance (J m^{-2} K^{-1})
n	Number of simulation steps
s	Sensitivity index for differential sensitivity analysis
R^2	Regression index

R. Pernetti (✉)
University of Pavia, Via Ferrata 1, 27100 Pavia, Italy
e-mail: roberta.pernetti@unipv.it

A. Prada
Faculty of Science and Tecnology, Free University of Bozen/Bolzano, piazza Università 5, 39100 Bolzano, Italy
e-mail: alessandro.prada@unibz.it

P. Baggio
Department of Civil Enviromental and Mechanical Engineering, University of Trento, via Mesiano 77, 38123 Trento, Italy
e-mail: paolo.baggio@unitn.it

A. Magrini (ed.), *Building Refurbishment for Energy Performance*, Green Energy and Technology, DOI: 10.1007/978-3-319-03074-6_5,

U	Thermal transmittance (W m^{-2} K^{-1})
O_j	Model response of the jth simulation

Greek Symbols

Λ	Thermal conductance (W m^{-2} K^{-1})
θ	Dry bulb temperature (K)

Subscripts

C	Cooling
f	Floor
H	Heating
I	Internal
r	Roof
sim	Simulated
set	Set point
w	Wall

1 Introduction

Energy simulation represents a useful tool to describe actual building behaviour; hence, it is applied not only in the design process but also in the post-occupancy analysis. In this case, the purpose is the evaluation of building actual energy efficiency in order to estimate the potential energy savings of existing constructions. In fact, the recast Energy Performance of Buildings Directive (EPBD [28] 2010/31/EU) highlights that residential and commercial buildings account for more than one-third of total annual energy consumption. Since significant energy savings can be achieved through energy conservation measures (ECM) for existing building stock, the importance of refurbishment has been growing. Consequently, simulations have been applied to the existing constructions to assess their energy performance and to define effective ECM. In this regard, dynamic energy simulation allows to understand the dynamic interactions between climate, building, occupants and energy systems. However, the large number of required inputs and parameters affects the reliability of dynamic simulation and significant discrepancies between predicted and real data could occur. Furthermore, the comparison between actual consumption and quasi-steady state prediction highlights important deviations. For these reasons, model calibration with monitored data is often appropriate in order to refine models and to develop more realistic energy-behaviour simulations.

Model calibration is widely used for commercial and office buildings analyses, which require the definition of complex transient simulations in order to design effective ECM. In fact, due to the large dimensions of these buildings and the operational daily schedules that vary with hourly interval, dynamic simulations are necessary to assess reliable energy behaviour.

Nevertheless, model calibration could be applied even for residential buildings. Quasi-steady state simulations usually require limited in situ measurements for model calibration, and they provide for the general behaviour of constructions (most of all with monthly intervals). Moreover, the input data are similar to the parameters employed in energy labelling, and the model calibration can be performed with utility bills. Consequently, this kind of evaluation is not time-consuming, and it guarantees for economical sustainability. Therefore, quasi-steady state simulations represent a useful tool for energy performance evaluation of both average- and small-dimension buildings.

On the other hand, a different approach is suitable for large-dimension houses, whose typology often constitutes social housing and it is widely spread in the suburbs of all the European cities. These constructions need a general refurbishment, most of all in terms of energy requirements, and they represent one of the strategic targets of EPBD [28] 2010/31/EU, because of the significant potential savings. In this regard, a large-scale evaluation as well as important investments are necessary for a general energy renewal. Thus, transient simulations can represent an important tool to plan effective ECM. Finally, time and economical costs due to dynamic models are sustainable in relation to these construction dimensions and to the potential energy savings.

Therefore, even in this case, model calibration is necessary in order to define an accurate model of the 'as-built' building and to design effective ECM.

According to these considerations, the large application field of model calibration requires operative procedures. A new European standard is going to be developed by CEN Technical Committee 89 (Working Group 14), and it will provide for calibration strategies and measurements of post-processing procedures for building energy models [24]. Currently, three protocols define general criteria and tolerance ranges for model calibration:

- International Performance Measurement and Verification Protocol (IPMVP [27]),
- Measurement and Verification (M&V Guidelines [31]);
- ASHRAE Guideline 14/2002: Measure of energy and demand savings [21].

However, none of these standards define an operative procedure to calibrate building models. In the literature, several studies deal with the model calibration issues using actual energy consumption either from in situ monitoring during the calibration period (e.g., [10, 12]) or from the analysis of monthly utility bills (e.g., [20]). Only a few works adopt the internal temperature as a calibration goal (e.g., [16]). In fact, this approach could be affected by a series of uncertainties and interactions with the indoor environment: occupant behaviour, internal gain and building equipment. Besides, the measurement of several variables can be an expensive and time-consuming activity. However, the model calibration using temperature as a control variable is the only viable procedure when no operating HVAC systems are present in a building.

Finally, the aim of model calibration is to minimize the discrepancies between the model and the real behaviour of buildings, therefore an extended procedure is necessary.

2 The Calibration Approach

Model calibration is an iterative process that aims to reduce the discrepancies between simulated and actual building energy behaviour, through the refinement of the model parameters.

In order to ensure the reproducibility of the calibration process and to reduce the uncertainties of model predictions, it is necessary to establish a reference procedure that defines the operative methodologies and the evaluation criteria of building properties. The calibration protocol presented and applied in this chapter is defined for existing buildings. It is compounded by several phases as shown in Fig. 1.

2.1 Preliminary Operations

The first step of a calibration process is the definition of scope and application field. In this chapter, a procedure for the calibration of existing building energy models is reported. During the early stage of the analysis, it is fundamental to check whether fuel measurement devices are available for the HVAC systems. Otherwise, the possibility of experimental equipment installation (for short- or long-term measures) should be investigated. Nevertheless, if a direct measure is not possible, building energy consumption can be derived from the utility bills, with a lower accuracy level respect to direct measurement.

2.2 General Data Gathering and Base Model Definition

The definition of a simulation model requires a large number of input data:

- *Building features*: dimensions and thermophysical properties of materials
- *HVAC system*: typology and technical features of the subsystem appliances, schedules and control strategies
- *Operating conditions*: internal gains due to lighting, equipments and people, zone occupancy and set point temperatures
- *Weather data*: dry bulb temperature, solar radiation, relative humidity and wind speed evaluated at the building location.

These data are used for the development of the base reference model. This model is defined using the real features and dimensions of building collected

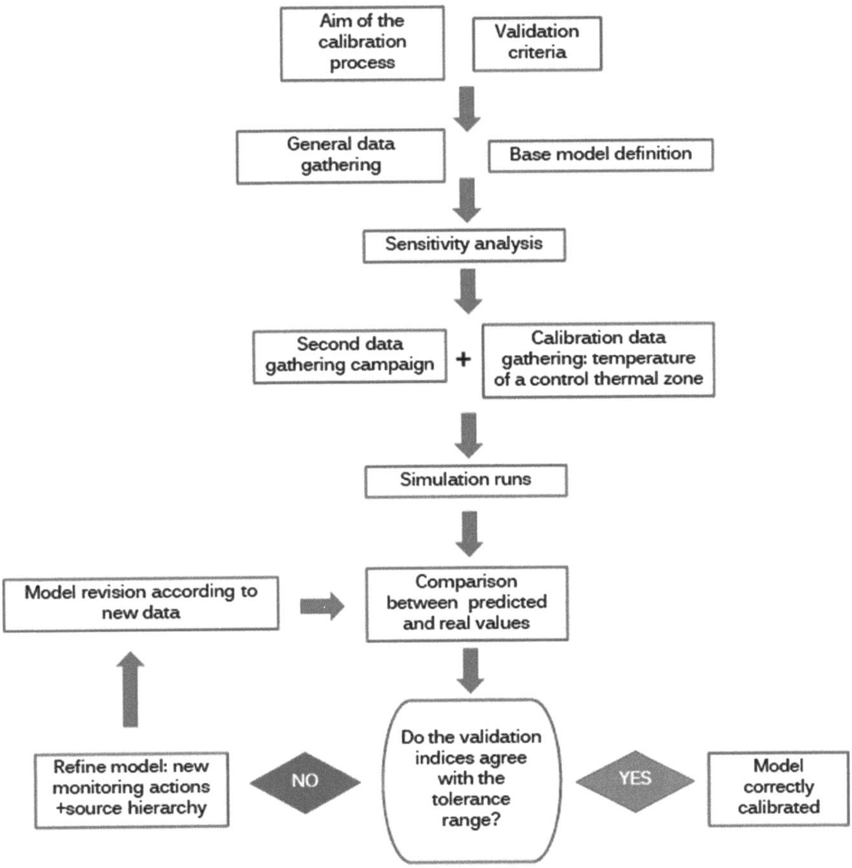

Fig. 1 Model calibration process

through a geometrical relief. Furthermore, for all the other parameters, standard reference values are used. This base model does not reproduce reliable energy behaviour of the building, but it represents the reference simulation for carrying out the sensitivity analysis (SA).

2.3 *Sensitivity Analysis*

The SA aims to evaluate the influence of input data on the dependant variables that, in case of building simulation, represent the energy behaviour of constructions (consumption and temperature trends).

The SA could be carried out with several methods, which are expounded in Sect. 4.

2.4 Second Data Gathering Campaign and Simulation Runs

During this phase of the calibration process, two categories of data are collected:

- Model input
- Control variables for calibration

The measurement of real values for the control variables for calibration allows to assess the reliability of the model during the validation step.

On the other hand, the gathering of model input for building simulation is highly complex due to the large amount of parameters that can affect the model results. Therefore, it is necessary to define strict criteria for the data selection. The SA highlights the most influent parameters and inputs that have to be investigated to obtain an effective simulation. Moreover, in order to refine the model, a source hierarchy has to be defined as explained in Sect. 5.

Finally, a series of simulations is carried out considering different inputs and boundary conditions.

2.5 Calibration Criteria

The discrepancies between real measured control parameters and model results have to be evaluated through given validation criteria, which assess the level of model calibration. In this phase, the indices and the tolerance range for the reliability assessment have to be defined.

2.6 Model Validation

In this phase, the model outputs are compared to the actual values of control variables for model calibration. If the results respect the calibration criteria, the model is correctly defined, otherwise it has to be refined by changing the model parameters or through a new data collection campaign, carried out according to the source hierarchy.

Finally, the calibrated model must be validated over a different measurement period respect to those used for the calibration phase. In fact, the calibration of the simulation model is a so-called inverse heat transfer problem, for which the uniqueness of the solution cannot be taken for granted [11].[1]

---

[1] "Inverse Heat Transfer Problems (IHTP) rely on temperature and/or heat flux measurements for the estimation of unknown quantities appearing in the analysis of physical problems in thermal engineering. As an example, inverse problems dealing with heat conduction have been generally associated with estimation of unknown boundary heat flux, by using temperature measurement taken below the boundary surface. Therefore, while in the classical direct heat conduction problem the cause (boundary heat flux) is given and the effect (temperature field in

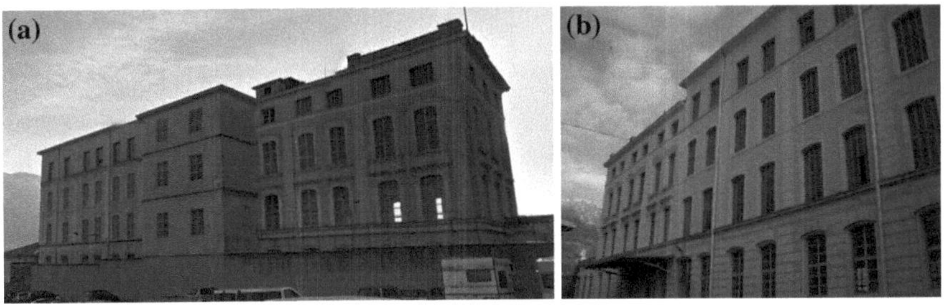

Fig. 2 **a** North-east facade. **b** North-west facade

3 Case Study

The case study is a historical manufacturing facility built in Rovereto in Northern Italy. The building, presently disused, was realized in 1854 as a storage construction for the tobacco processing (Fig. 2).

It has an overall surface of 3,650 m^2, four levels, one basement and a flat roof with a black coating (absorption coefficient roughly equal to 0.9).

The building has a concrete structure (beam and pillar) and a massive envelope made of stone and bricks, whose thickness ranges from 90 cm of the underground floor to 65 cm of the third floor. Except for the underground level, which is characterized by basement windows, the envelope has a homogeneous ratio of glazing over opaque surface equal to 30 %.

The windows have timber frames with single glazing. During the relief, several leakages in the glass elements were detected, with the consequence of significant infiltration rates.

The building is now disused, and therefore, it has no operating HVAC systems.

Considering the high thermal capacitance of the internal walls, each room is modelled as a single thermal zone, as shown in Figs. 3 and 4.

The envelope material properties are unknown, and there are no available design documentations. Therefore, according to the building features, the construction year, the structure and the thickness, and standard compositions were extracted from the Appendix A of the Italian technical specification UNI/TS 11300-1 [30] (Table 1).

4 Sensitivity Analysis of Building Energy Model

The detailed modelling of the building and HVAC energy performances leads to careful study of the interactions between the envelope, the occupants, internal loads and energy systems. However, the increasing detail of the models requires a

(Footnote 1 continued)
the body) is determined, the inverse problem involves the estimation of the cause from the knowledge of the effect" (Ozisik and Orlande 2000).

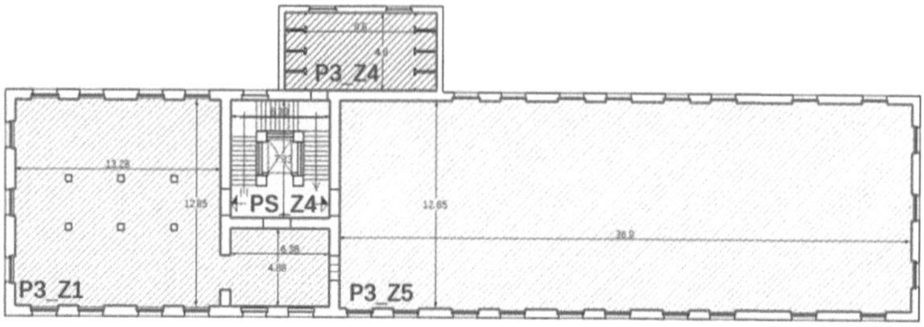

Fig. 3 Thermal zones

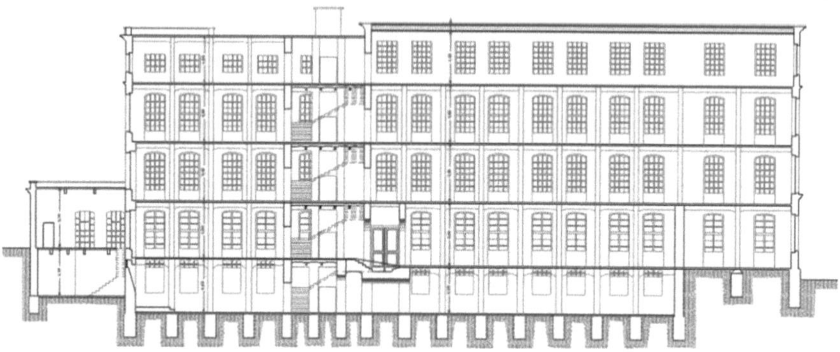

Fig. 4 Building section

greater number of input data that, if not properly investigated, can undermine the reliability of results. Furthermore, the relative influence of stochastic variations in building energy needs increases for low energy constructions. On the other hand, the input data do not affect in the same way the model predictions and, in this sense, it is important to carefully assess the sensitivity of the model to the input parameters.

A practical definition is that SA is an answer to the questions 'To what extent simulation predictions are reliable if input data are affected by uncertainty or are known with certain accuracy?' or 'To what extent the accuracy and the precision of model previsions improve if the knowledge of the input data is increased?'

A more precise definition is provided by Kioutsioukis et al. [5], which defines the SA as 'the study of how uncertainty in the output of a model (numerical or otherwise) can be apportioned to different sources of uncertainty in the model input'.

The main application of the SA in the calibration of the building energy model thus becomes the identification of the critical inputs for the results reliability. This knowledge allows to establish priorities and to limit the experimental activities for input data measurements. In this way, it will be possible to minimize the in situ measures, which can become time-consuming activities.

Table 1 Envelope composition

Brick and stone wall

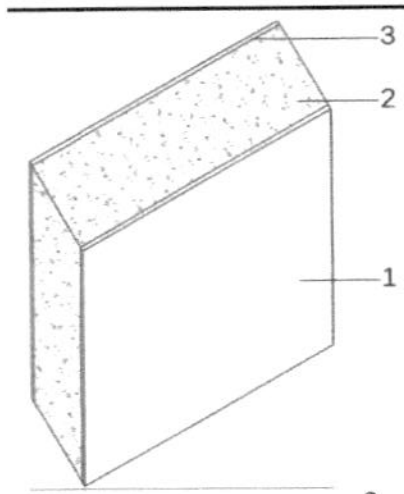

Layer	d (cm)	ρ (kg/m^3)	λ [W/(m K)]	R (m^2 K/W)
1 Internal plaster	2	1,400	0.700	–
2 Brick and stones	30–65	1,500	0.900	–
3 External plaster	2	1,800	0.900	–

Ground floor

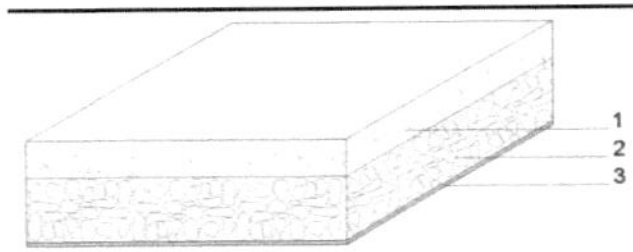

Layer	d (cm)	ρ (kg/m^3)	λ [W/(m K)]	R (m^2 K/W)
1 Mortar	3	2,000	1.470	–
2 Concrete	10	2,000	1.400	–
3 Stones	20-40	1,700	1.160	–

Intermediate floor

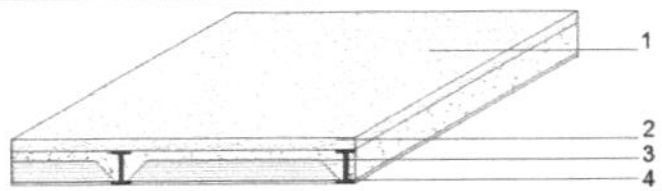

Layer	d (cm)	ρ (kg/m^3)	λ [W/(m K)]	R (m^2 K/W)
1 Mortar	2	2,000	1.400	–
2 Light concrete	9	400	0.150	–
3 Hollow brick	6	600	–	0.250
4 Plaster	2	1,800	0.900	–

Plain roof

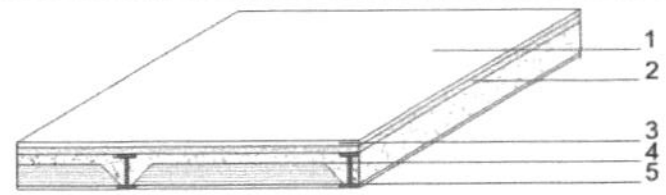

Layer	d (cm)	ρ (kg/m^3)	λ [W/(m K)]	R (m^2 K/W)
1 Coating	1.5	1,200	0.170	–
2 Mortar	2	2,000	1.400	–
3 Light concrete	9	400	0.150	–
4 Hollow brick	6	600	–	0.250
5 Plaster	2	1,800	0.900	–

Although in the literature there are several techniques for the SA [15], this chapter only describes a few simple techniques applied to building energy models, by highlighting the advantages and application limits.

As reported by Macdonald [8], the SA can be divided in two categories:

- *External methods*: where a sample of input is generated and, subsequently, the deterministic numerical model is executed for each input.
- *Internal methods*: that directly evaluates the output distribution from the uncertain inputs and from the differential equations of the mathematical model.

The external procedures are also divided in two branches, i.e., local and global. In local methods, the output uncertainty is evaluated with respect to changes around a specific value of individual parameters, whereas global methods quantify the output variation along a specific range of input variability.

The two approaches adopted in this chapter belong to the local methods based on the derivatives techniques, which have the advantage of being very efficient computationally with respect to global sensitivity methods such as Monte Carlo technique. In fact, derivative methods require only few simulations compared with the Monte Carlo methods. However, the main drawback is the limited suitability for models of unknown linearity. In fact, derivative methods provide punctual information about the model sensitivity (i.e., local method) and they do not allow an extrapolation of the results to the rest of the input space. In non-linear models, therefore, the impossibility to extend the results to other input values, with respect to those used for the estimation of the local sensitivity index, requires a reliable estimation of the input and of its expected variability range. These data must, therefore, be derived from the experience of the energy modeller thus to assure useful results from the SA.

The two techniques adopted in this chapter (local techniques based on the derivatives) do not represent the most detailed ones, but rather they could be easily integrated in energy simulations and, therefore, they are more useful for the application to real cases. Indeed, the ratio of an output O over an input I can be thought as a mathematical definition of the sensitivity of output model with respect to input variability.

4.1 Differential Sensitivity Analysis

The differential sensitivity analysis (DSA) is the backbone of local methods, and it works by perturbing an input data around the mean value while all the other parameters remain fixed [7]. For each perturbed value, the numerical simulation is carried out and the model response is calculated. Due to its robustness and simplicity, the DSA is the most diffuse method for a local uncertainty evaluation. The effects of an uncertain parameter are estimated by comparing the results of these simulations against those with unperturbed inputs. Consequently, a sensitivity index (s) of the model prediction to the uncertain parameter is defined as:

$$s = \frac{\Delta O}{\Delta I} \tag{1}$$

where ΔO is the difference between the output with perturbed input with respect to deterministic model as well as ΔI is the perturbation of input.

Since the sensitivity index depends on both the input and output dimensions, Lam and Hui [6] proposed a dimensionless index ($s_{\%}$) defined as:

$$s_{\%} = \frac{\Delta O / O_{\text{base}}}{\Delta I / I_{\text{base}}} \tag{2}$$

where the numerator and denominator report, respectively, the percentage changes of output and input of the model.

The main weakness of this calculation procedure is that it assumes the perfect independency among all parameters. Consequently, only the elementary effect of each parameter could be computed while the combined effects can be estimated by a superposition in linear problems.

4.2 Factorial Methods

With the aim of overcoming DSA issues, the factorial method (FM) is also used in SA. The FM is a further development of the DSA approach that includes the interactions between parameters and, consequently, it permits the estimation of the higher-order effects. In this procedure, all the uncertain parameters are perturbed simultaneously around their mean values.

Usually, two different perturbation levels are considered for each parameter by imposing a low and high level (Table 2). In general, k parameters would require 2^k simulations to generate all combinations for a full factorial simulation plan. These combinations represent the corners of the k-dimensional hypercube. Hence, the drawback of this technique is the number of simulations required that increase faster with the number of inputs.

If the factorial design of simulation aims to determine the model behaviour at a grid of locations inside the hypercube, more than 2 levels for each parameter are required. In this case, the total number of simulations becomes l^k, where l is the number of perturbation levels.

Based on the model predictions, the first-order effects can be calculated as:

$$F_A = \frac{(O_2 + O_4 + O_6 + O_8) - (O_1 + O_3 + O_5 + O_7)}{4} \tag{3}$$

$$F_B = \frac{(O_3 + O_4 + O_7 + O_8) - (O_1 + O_2 + O_5 + O_6)}{4} \tag{4}$$

Table 2 Factorial design for three inputs with two perturbation levels

Number	A	B	C	A–B	A–C	B–C	A–B–C
1	−	−	−	+	+	+	−
2	+	−	−	−	−	+	+
3	−	+	−	−	+	−	+
4	+	+	−	+	−	−	−
5	−	−	+	+	−	−	+
6	+	−	+	−	+	−	−
7	−	+	+	−	−	+	−
8	+	+	+	+	+	+	+

$$F_C = \frac{(O_5 + O_6 + O_7 + O_8) - (O_1 + O_2 + O_3 + O_4)}{4} \tag{5}$$

where F_i is the first-order effect due to the ith input and O_j represents output of the jth simulation run.

The higher-order effects can be computed starting from the simulation runs. In this regard, the signs to be used in the following equations are obtained by multiplying the sign reported in (Table 2)

$$F_{A-B} = \frac{(O_1 + O_4 + O_5 + O_8) - (O_2 + O_3 + O_6 + O_7)}{4} \tag{6}$$

$$F_{A-C} = \frac{(O_1 + O_3 + O_6 + O_8) - (O_2 + O_4 + O_5 + O_7)}{4} \tag{7}$$

$$F_{B-C} = \frac{(O_1 + O_2 + O_7 + O_8) - (O_3 + O_4 + O_5 + O_6)}{4} \tag{8}$$

$$F_{A-B-C} = \frac{(O_2 + O_3 + O_5 + O_8) - (O_1 + O_4 + O_6 + O_7)}{4} \tag{9}$$

where F_{A-B} is the second-order effect due to the Ath and B^{th} inputs.

In the next section, the SA on the test case presented in Sect. 3 is reported. In particular, this analysis is carried on as a preliminary investigation of the hourly energy simulation model with the purpose of understanding the extent to which each parameter can affect the calibration procedure. Starting from this information, some of these inputs are further investigated by experimental activity or using a higher hierarchy source.

4.3 Example

In this section, the results of the SA on the case study described in Sect. 3 are presented. Since in the test case there is no energy systems, the dependent variables are related to the air temperature of the control thermal zone (i.e., P3_Z1).

Besides, four different variables are adopted with the purpose of analysing some indexes closely related to system sizing and to the estimation of energy needs. In particular, the variables adopted are as follows:

- Minimum zone air temperature ($t_{\min}$)
- Maximum zone air temperature ($t_{\max}$)
- Zone Heating Degree Hour (HDH_{18})
- Zone Cooling Degree Hour (CDH_{26})

Heating and Cooling Zone Degree Hour indicate the sum of hourly difference between internal set point temperature (i.e., 18 °C for heating and 26 °C for cooling) and the simulated values of air temperature in the thermal zone.

$$\mathrm{HDH}_{18} = \sum_{i=1}^{n} \left(\vartheta_{i,H,\mathrm{set}} - \vartheta_{i,H,\mathrm{sim}}\right) \tag{10}$$

$$\mathrm{CDH}_{26} = \sum_{i=1}^{n} \left(\vartheta_{i,C,\mathrm{sim}} - \vartheta_{i,C,\mathrm{set}}\right) \tag{11}$$

These indices are adopted because they are proportional to the heating and cooling demand, as well as minimum and maximum temperature are closely related to the required size of energy system.

Starting from this point, a SA is carried out with a local external approach using both the DSA and the FM approaches. Based on some analyses about the uncertainty level of some parameters, the SA is performed by perturbing the following:

- External envelope air tightness expressed as an airflow for pressure difference of 4 Pa (Q4 Pa)
- Roof thermal transmittance (U_{roof})
- Wall thermal transmittance (U_{wall})
- Intermediate Floor thermal transmittance (U_{floor})
- Roof thermal specific capacitance (κ_{roof})
- Roof thermal specific capacitance (κ_{roof})
- Wall thermal specific capacitance (κ_{wall})
- Floor thermal specific capacitance (κ_{floor})
- g-value for glazing systems (g-value)

Starting from the base model described in Sect. 3, each parameter is perturbed by applying a ± 10 % variation in the original value.

Figure 5 shows the $s_{\%}$ sensitivity index, respectively, for HDH_{18} and CDH_{26} (Fig. 5a) and for minimum and maximum zone air temperature (Fig. 5b).

Note that for CDH_{26}, g-value and roof thermal transmittance are the most influent parameters. Besides, for these variables, the indices have a positive sign that indicates a direct correlation. The greater the input values the higher the CDH_{26} and, consequently, the cooling demand. The other indices are negative but the magnitudes of sensitivity index are close to zero. The graphs highlight the role

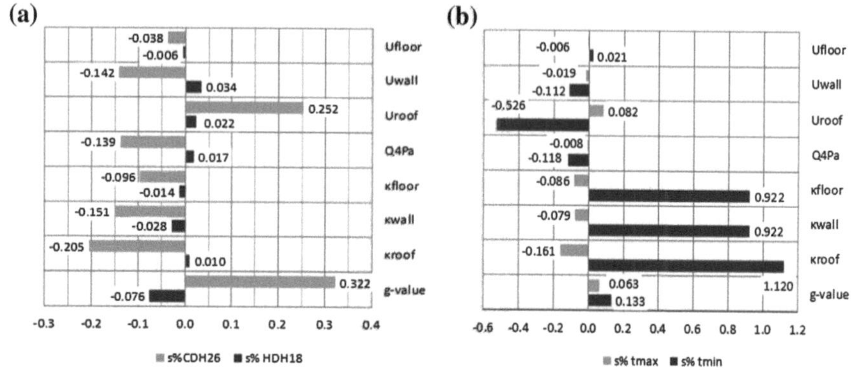

Fig. 5 $S_{\%}$ for zone heating and cooling degree hour and for minimum and maximum zone air temperature

of thermal capacitance both of wall and roof in keeping down the cooling demand. Besides, it is interesting to note the negative correlation between CDH_{26} and the wall thermal transmittance, it means that, for the test case, the night heat losses prevail on the inward heat losses.

On the other hand, Fig. 5b shows that thermal capacitance of the envelope strongly affects both minimum and maximum temperature for the investigated test case. Lower magnitude is registered for the other parameters and in particular it is interesting to note the low effects of *g*-values with respect to envelope capacitance.

These graphs show to which parameters the model simulation is more sensitive. Additionally, the factorial analysis method provides information on the high-order effects and then on the interactions among different parameters. The main difference is hence the simultaneous perturbation of the parameters aiming to discover the possible synergistic effects of variable perturbations. Besides, in order to compare the results both for degree hour indices and for internal peak temperatures, also the relative factorial factors are used. These indices are calculated by dividing the results of the Eqs. (6–9) for the unperturbed energy demand. The results obtained (Table 3) are consistent with the DSA.

Regarding first-order effects, the FM confirms that HDH_{18} and CDH_{26} are less affected by thermal capacitance of floor, whose index is of an order of magnitude lower than $F_{\kappa\text{wall}}$ and $F_{g\text{-value}}$.

The results of factorial analysis show weak second-order effects, and the link among variables has generally a negative sign, which means that there is not a synergistic effect. Therefore, the assumption of perfect independent variables of the DSA approach has been proved for this particular energy model. Starting from these considerations, the model can be refined by further investigating the most sensitive parameters. Moreover, also in model calibration, particular attention will be paid to the investigation of the parameter with the highest sensitivity indexes.

Table 3 Relative factorial indexes

Effect	CDH_{26}	HDH_{18}	t_{max}	t_{min}
$F_{\kappa f}$	−0.0038	−0.0009	−0.0015	0.0221
$F_{\kappa w}$	−0.0302	−0.0042	−0.0145	0.1827
$\boldsymbol{F}_g$	0.0715	−0.0169	0.0138	0.0314
$\boldsymbol{F}_{\kappa f-\kappa w}$	0.0004	0.0001	0.0000	0.0000
$\boldsymbol{F}_{\kappa f-g}$	0.0000	0.0000	−0.0001	0.0000
$\boldsymbol{F}_{\kappa w-g}$	−0.0007	−0.0001	−0.0004	0.0018
$F_{\kappa w \boldsymbol{kf} g}$	0.0000	0.0000	0.0000	0.0000

5 Collecting Input Data

The data gathering is a critical and complex step. Because of the large amount of parameters and inputs that affect the model, the number of missing data could be significant. Therefore, each experimental campaign is affected by a certain level of approximation, whose extent has to be established at preliminary level according to the general accuracy needed for the model.

Moreover, the data collection depends on the kind of simulation to be carried out: transient simulations reproduce detailed results, but in situ measurements are necessary to define the trends of parameters along the simulation period. On the other hand, quasi-steady state models use monthly data with a lower level of reliability in the simulation of real building behaviour.

Raftery et al. [13, 14] propose the definition of a source hierarchy for each calibration: it assigns a level of accuracy to an input data according to the reliability of the source evidence. In particular, a general hierarchy for a building simulation, as indicated also in the Guidelines ASHRAE 14 [21], is composed by:

- Direct sources
  - Long-term monitoring (>6 months)
  - Short-term monitoring
  - Spot measurement
  - Direct relief of the internal environment
  - User interview
- Indirect sources
  - Design project and documentation
  - Technical sheet of materials and operating manual of the HVAC system
- Standard sources
  - Technical standards
  - Standard guidelines and reference catalogues

In case of low-level sources and indirect investigations, spot measurements and visible inspections are necessary to verify the reliability of documentation.

Standard references represent useful tools to integrate the building model data, but they provide for general information not related to the specific building. Hence, they could be a potential discrepancy source for the building simulation.

Moreover, in a calibration process, the source hierarchy represents an important reference to refine the building simulation.

In the further sections, some measurement techniques and example of data gathering are reported.

5.1 Experimental Data for Model Calibration

In order to calibrate the building energy model, the simulation predictions have to be validated and verified against experimental measurements. Generally, the protocols and regulations indicate the actual fuel or power consumption of energy system as comparative data for model calibration. However, it is not always possible to trace the actual energy consumption, as for disused buildings or in constructions without energy systems. Therefore, in these cases, the indoor air temperature as well as the envelope surfaces temperatures may be employed as control variables for the model calibration.

The experimental data collection becomes then a key aspect in the calibration process of an energy model and, at the same time, it can be one of the most complex and expensive topic in the energy analysis. For these reasons, every in situ measurement must be based on a design of experiment established in early stage of the analysis. This document has to define the measurements with respect to the energy simulation requirements and, in particular, for the inputs highlighted by SA or for data that the energy modeller believes affected by epistemic uncertainty. The design of experiment has to define the instruments' requirements in terms of accuracy and precision, the maximum sample rate, the data quality control and assurance, the expected range of experimental data, the procedures for detection and management of outliers and missing data.

The following paragraphs will discuss some of the most frequents experimental activities carried out for the calibration of building energy models.

5.2 Thermal Conductance Measures

The measurement of envelope thermal conductance is to a large extent one of the key measures for the energy simulation refinement. Indeed, in constructive elements in which the materials are unknown, the uncertainty in the estimation of the overall conductance can be propagated through the energy model.

The standard ISO 9869:1994 [29] defines the main aspects concerning the experimental approach: it indicates the equipment to be used, methods of measurement and the quality assurance and the post-processing techniques.

Specifically, the ISO 9869 methods are valid for opaque elements characterized by normal heat flow with negligible lateral heat flows.

The measurement has to be carried out by means of one heat flow meter on the internal side (i.e., the side adjacent to the most stable temperature) and two thermometers positioned both on the internal and external surface of the wall. The heat flow meter is a thermopile of known thermal conductivity that measures the temperature difference between the two faces of the plate. Moreover, a thin silicon paste is added between heat flow meter and wall surface aiming to decrease the contact thermal resistance.

The standard sets out the measurement conditions to be met in order to ensure the reliability of results. In addition, the sampling rate and the test duration are also defined as a function of the wall characteristics, temperature trends and the post-processing method. The minimum test duration has to be greater than 72 h. However, if the temperature or the heat flux has a variable trend over time, the duration must be extended for a period longer than 7 days, until stable results are reached.

Two different post-processing techniques are proposed in the international standard, i.e., the average method and the dynamic analysis method. The average method computes the thermal conductance of the building element as the ratio of the mean density of heat flow rate over the mean temperature difference, as reported in Eq. (12).

$$\Lambda = \frac{\sum_{j=1}^{N} (q_j)}{\sum_{j=1}^{N} (\theta_{\mathrm{si},j} - \theta_{\mathrm{se},j})} \tag{12}$$

If the conductance value is estimated after each measurement, it converges to an asymptotical value (Fig. 6b). On the other hand, with the purpose of ensuring a faster solution convergence, the standard suggests the use of dynamic analysis method. Starting from the temperatures and the heat flow collected, at each time the heat flow rate can be obtained from (13):

$$\begin{aligned} q_i = {} & \Lambda(\theta_{\mathrm{si},i} - \theta_{\mathrm{se},i}) + K_1 \dot{\theta}_{\mathrm{si},i} - K_2 \dot{\theta}_{\mathrm{se},i} \\ & + \sum_g P_g \sum_{j=i-p}^{i-1} \theta_{\mathrm{si},j}(1-\beta_g)\beta_g(i-j) \\ & + \sum_g Q_g \sum_{j=i-p}^{i-1} \theta_{\mathrm{se},j}(1-\beta_g)\beta_g(i-j) \end{aligned} \tag{13}$$

where K_1 and K_2 and P_g and Q_g are dynamic characteristics depending on the g time constants τ_g. The variables β_g are exponential functions of the time constants. Since Eq. (13) is a function of the $2g + 3$ unknowns, at least $2g + 3$ data points are needed for the solution of the linear systems. However, with the purpose of eliminating stochastic variations, an over determined system of M equations is usually solved by means of least square fit. The accuracy of the outcome of this

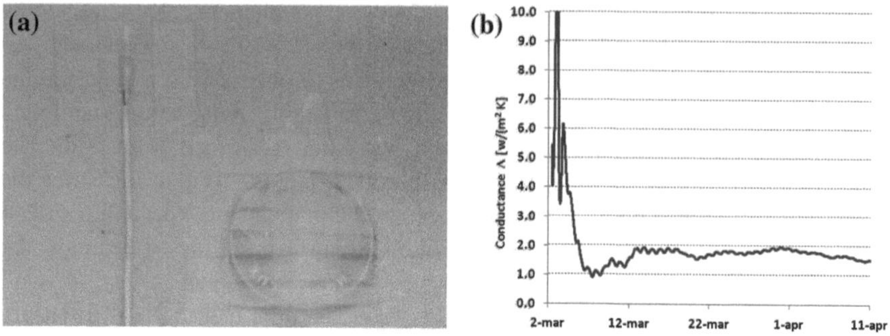

Fig. 6 Thermal transmittance evaluation: **a** Heat flux meter and Pt100 position. **b** Thermal conductance evaluated by Standard ISO 6946

method is a function of the number of data analysed (*N*), of the choice of the number (*g*) and ratio (*r*) of time constants and of the dimension (*M*) of the linear system [2].

5.2.1 Example

The two post-processing methods are applied on the measures collected to a wall of the test case presented in Sect. 3. In particular, due to the massive wall and to the high temperature variation also at the internal side of the component, the measurement period is extended for more than a month in order to ensure the convergence of the solution. In Fig. 6a, the positions of the thermoresistance (Pt100) and of the heat flow meter both on the outer side of the wall are shown.

The first post-processing technique is the average methods applied to the measurement period that met the conditions:

- $\Lambda_{END} - \Lambda_{24} < 5\ \%\ \Lambda_{END}$;
- $\Lambda_{END} - \Lambda_{2/3} < 5\ \%\ \Lambda_{END}$;

where Λ_{END} is the final thermal conductance, Λ_{24} and $\Lambda_{2/3}$ are respectively the conductance obtained at 24 h before the end of the monitoring and at 2/3 of the same.

The second result is obtained applying the dynamic method described in the previous section (Eq. 13) (Table 4).

5.3 *Weather Data for Model Calibration*

Weather data represent one of the main external forces driving the building energy behaviour. Therefore, it is important to refer both the actual energy consumption and the indoor environmental variables to the actual weather conditions [17].

Table 4 Conductance values obtained by two post-processing methods

Average method	Dynamic analysis method
1.552 W/(m^2 K)	1.439 W/(m^2 K)

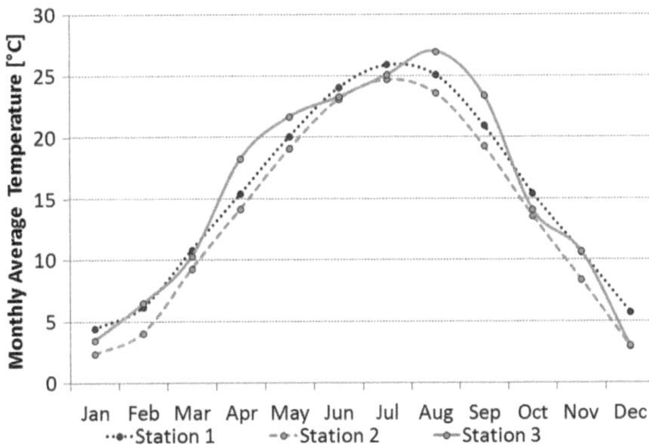

Fig. 7 Monthly average temperatures in Pavia meteorological stations

These climate conditions are then used as boundary conditions for the energy model in order to obtain consistent result with respect to measurements.

Furthermore, it should be stressed that weather data are not independent variables but rather a single set of cross-correlated measures collected at the same site location [4]. Additionally, the sample rate must be consistent with the model undertaken and, therefore, with the phenomena variability coupled with building or system response.

The data required for calibration can be obtained from meteorological stations located close to the building or directly measured in situ through the installation of a weather station. In fact, the suitability of data collected at meteorological stations cannot be taken for granted since the representativeness of local conditions are not assured owing to local climate effects such as urban heat island, urban canyons, or orographic influences [3].

As an example, Fig. 7 shows the dry bulb temperature collected in 2011 for the city of Pavia (Italy). Three meteorological stations are available in the province as shown in Table 5.

The trends of average monthly temperatures show noticeable deviations between the data recorded in different parts of the same city. This spread is mainly related to the effects of urban heat island and, consequently, to the interactions with the surrounding of meteorological station.

Therefore, it is important to correctly verify the data prior to use it in the model calibration process. In this regard, it is useful to distinguish two different stages of validation process. As defined by the guide WMO n. 8 [32], they are the quality assurance and quality control.

Table 5 Weather data stations

Station	Context	Coordinate
1	Urban	45° 11′ 42.56″ N, 9° 9′ 48.42″ E, 79 m asl
2	Suburban	45° 10′ 51.331″ N, 9° 8′ 48.053″ E, 55 m asl
3	Urban (city centre)	45° 11′ 8.931″ N, 9° 9′ 28.463″ E, 87 m asl

1. *Quality assurances* are all the planned and systematic activities implemented within the quality system to provide adequate confidence that an entity will fulfil requirements for quality. The primary objective of the quality assurance system is to ensure that data are consistent, meet the data quality objectives and are supported by comprehensive description of methodology.
2. *Quality controls* are the operational techniques and activities that are used to fulfil requirements for quality. The primary purpose of quality control of observational data is missing data detection, error detection and possible error corrections in order to ensure the highest possible reasonable standard of accuracy for the optimum use of these data by all possible users.

These two definitions clearly indicate that, when data are obtained from meteorological stations, the quality assurance activities are the responsibility of the owner of the station, while the verification of raw data (quality control) can be charged to the end user. On the other hand, the energy modeller has to ensure the quality assurance in case of data obtained from an in situ weather station specifically installed for the purpose.

The quality control activities aim to discover the outliers and the unphysical data. In this regard, the WMO guide suggests these checks [1]:

- values exceeding more than 50 % the 1st and 99th percentile are deleted;
- temperature derivatives higher than 4 K/h are not physical;
- values repeated for more than five times for temperature, solar radiation and wind velocity are anomalous data;
- values repeated for more than five times for relative humidity are anomalous if lower than the 75th percentiles;
- global solar radiation higher than solar constant are eliminated as well as radiation before sunrise or after sunset;
- negative values of wind velocity, solar radiation and relative humidity as well as relative humidity higher than 100 % are unphysical.

The data that do not meet these requirements are deleted and treated as missing values in the interpolation phase. In addition, it is generally assumed a threshold of 25 % [23] of missing data in a specific measurement period to prevent that the excessive interpolation leads to the use of a synthetic dataset.

The use of linear interpolation is not always the best solution for data filling. In fact, the use of linear interpolation for temperature, relative humidity and wind velocity can be accepted for short period of missing data [12]. For longer period and for solar radiation, the linear interpolation cannot adhere to the natural

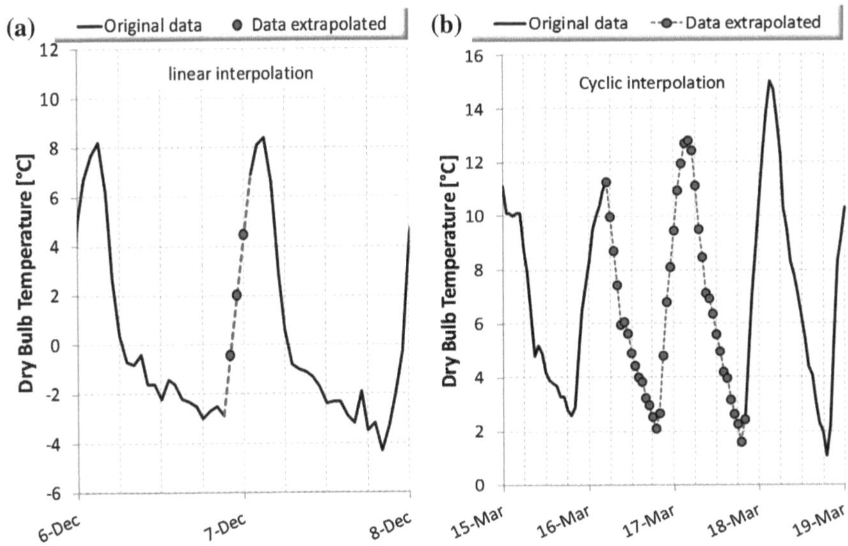

Fig. 8 Weather data interpolation: **a** Linear interpolation. **b** Cyclic interpolation

variation in the weather data. Therefore, in these cases, a cyclic interpolation is generally used (Fig. 8).

Lastly, it is important to remember that the weather data must comply with the energy model requirements. For example, if the energy analysis is performed by means of an hourly dynamic simulation, the main weather data required are the following [18]:

- Global horizontal irradiance: is the total amount of direct and diffuse solar radiation received on a horizontal surface during the 60-min period ending at the time stamp
- Dry bulb temperature: is the dry bulb temperature at the time indicated (instantaneous value)
- Relative humidity: is the relative humidity at the time indicated (instantaneous value)
- Wind velocity: is the wind speed at the time indicated (instantaneous value).

Therefore, particular attention should be paid when data collected in meteorological station are used. In fact, weather variables are routinely sampled with temporal frequencies of 10–15 min, while the published data are generally hourly average values. Therefore, the instantaneous values must be requested for temperature, relative humidity and wind speed whereas the adopted average techniques has to be investigated for the solar radiation values. In fact, there are no widely accepted standard and, consequently, each institution uses its own convention such as a forward, backward or centred average (Table 6).

Table 6 Evaluation of hourly average

Backward average			Forward average		
Measurements		Time stamp	Measurements		Time stamp
06:00	⇨	06:00	06:15	⇨	07:00
06:15			06:30		
06:30			06:45		
06:45			07:00		

6 Evaluation of Actual Energy Consumption

The actual fuel consumption is a reference for the calibration that allows to compare the real behaviour with the model results in terms of energy demand of the energy systems. This section describes some methods for the gathering of data consumption and for the data post-processing.

In particular, the actual energy needs are evaluated basing on the amount of fuel (by volume or weight) converted in energy through multiplication by the lower heating value according to the following relationship:

$$Q_{\text{real}} = V_{\text{fuel}} \cdot \text{L.H.V.} \tag{14}$$

where

- Q_{real} → actual energy needs
- V_{fuel} → fuel amount di
- L.H.V. → lower heating value[2] (reference values are reported Table 7).

The total energy consumption of a building consists in several components:

$$C_{\text{coll}} = C_h + C_W + C_{\text{cook}} \tag{15}$$

where

- Co_{coll} → real gathered consumption
- Co_h → heating $\text{Co}_h = \text{Co}_h = 0$ (during summer season)
- Co_W domestic hot water production
- Co_{cook} → for cooking

In Table 8 reference values for cooking energy needs for residential buildings are presented; these terms can be assumed as constant during all the monitoring period.

[2] Generally, only the H.H.V. (higher heating value) is reported in the utility bills. Nevertheless, L.H.V. is required to obtain effective energy consumption. Therefore LHV has to be estimated starting from the HHV.

Table 7 Reference lower heating values for common fuels

Fuel	Low heating value (L.H.V.)
Methane	9.940 kWh/Nm^3
Propane	28.988 kWh/Nm^3
Butane	36.779 kWh/Nm^3
Diesel fuel	11.870 kWh/kg

Table 8 Reference coking energy needs

Plan surface (m^2)	Specific energy need (kWh/G)
<50	4
50 m^2 < Surface < 120 m^2	5
>120 m^2	6

In addition, it is necessary to divide the energy needs for heating respect to domestic hot water requirements, which can be deducted through summer measurements, (after the cooking contribute subtraction).

In fact, energy needs for domestic hot water are assumed to be fixed throughout the year with an adequate level of approximation, if thermal solar systems are not installed. Hence the summer DHW consumptions can be used even for winter.

Another method to deduce domestic hot water, cooking and plug loads is based on the correlation between external temperature and energy consumption during the operational period of HVAC system [20]. In fact, when the load line becomes parallel with the x-axis, fuel consumptions are not affected by external temperature and the constant value represents the energy needs for the production of domestic hot water and for cooking. The values for a reference year could be represented in a graph in association with monthly temperatures: the lower the external temperature, the higher the heating consumption and vice versa. When there are no fuel consumption caused by domestic hot water (electrical production), the constant value is close to zero. Also the electrical requirements trend could be expressed with respect to the monthly external temperature. The slope increases according to the temperatures owing to the cooling system consumptions. In this regard, an horizontal line represents the sum of domestic hot water, lighting and plug loads.

6.1 Case Study: Utility Bills Analysis

The case study is a residential detached building with a floor surface of 120 m^2 heated by means of a combined boiler (heating and domestic hot water). The actual fuel consumption, collected during a whole year, is reported in Table 9.

Table 9 Real fuel demand of the case study

Reference period of the utility bills	Number of days	Fuel consumption (m^3)	Energy needs (kWh)
January	30	660.00	6,560.4
February–March	59	1,178.00	11,709.32
April–May	60	425.00	4,224.5
June–August	91	169.00	1,679.86
September–November	90	1,274.00	12,663.56
December	30	213.00	2,117.22

Table 10 Real energy needs (without cooking contribution)

Reference period of the utility bills	Number of days	Energy needs (kWh)	Cooking energy needs (kWh)	Net energy needs (kWh)
January	30	6,560.40	180	6,290.94
February–March	59	11,709.32	354	10,361.32
April–May	60	4,224.50	360	3,745.22
June–August	91	1,679.00	546	1,133.86
September–November	90	12,663.56	540	11,139.50
December	30	2,117.22	180	1,957.10

Utility bills report the fuel consumption, which can be converted into energy need through the Lower Heat Value that, for methane, is roughly equal to 9.940 kWh/m^3:

$$\text{Co(kWh)} = \text{Co}(\text{m}^3)\,(\text{m}^3) \times 9.940\ \text{kWh/m}^3$$

This equation has been applied to obtain the values in the fourth column (Table 9). Nevertheless, cooking energy needs and the contribution for domestic hot water have to be separated from the total consumption. Considering the floor surface and the reference values reported in Table 8, the energy requirements for cooking accounts for 6 kWh/day (Table 10).

The last contribution to be separated from the total consumption is the energy required for domestic hot water preparation. Considering the average energy consumption registered from period between June and August, the daily energy needs are evaluated as:

$$\frac{1,679\ \text{kWh}}{91\ \text{g}} = 12.46\ \text{kWh/day}$$

Extending this value to the whole period, the results showed in Fig. 9 are obtained.

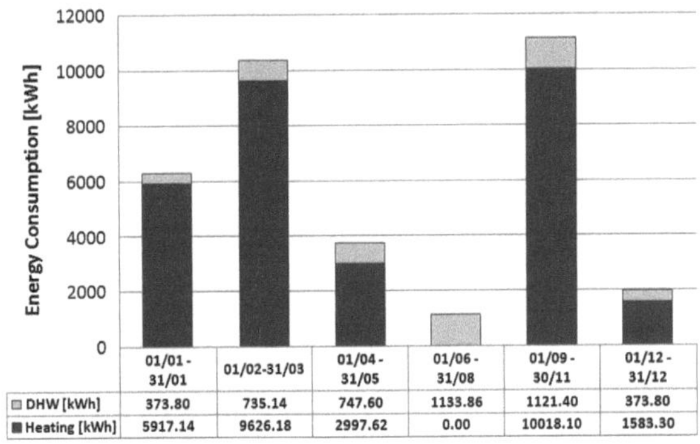

Fig. 9 Heating and domestic hot water consumption

6.2 Source for Real Consumption Monitoring

According to the fuel of the HVAC system, different sources of data are necessary to evaluate actual consumption.

For instance, in case of network gas systems, monitoring device are installed to evaluate the consumption and define the amount of the utility bills.

There are two main methods for the data record:

- *Indirect*
- *Direct*

In the first case, real consumption is estimated through the values on the utility bills in relation to the monitoring period. Nevertheless, these amounts could be determined by a statistical evaluation according to the previous consumption (estimate values). Therefore, they cannot reproduce the real energy behaviour of the building, but they show the historical trends of fuel supply. Hence, in order to calibrate the model, the effective energy needs have to be adopted.

On the other hand, the direct method is featured by a relief of the measurement device during the monitoring period. The finer the gathering interval, the more accurate the calibration.

In case of HVAC system fuelled by a storage volume with a measurement device, the model can be calibrated considering the instrument error. If the storage has no counter, the fuel consumption can be estimated through the following equation:

$$\mathrm{CQ} = (\mathrm{CQ}_I - \mathrm{CQ}_F) + \mathrm{CQ}_A \tag{16}$$

where

- $Q \rightarrow$ quantity in the storage
- Q_I initial quantity
- Q_F final quantity
- Q_A supplied quantity during the monitored period.

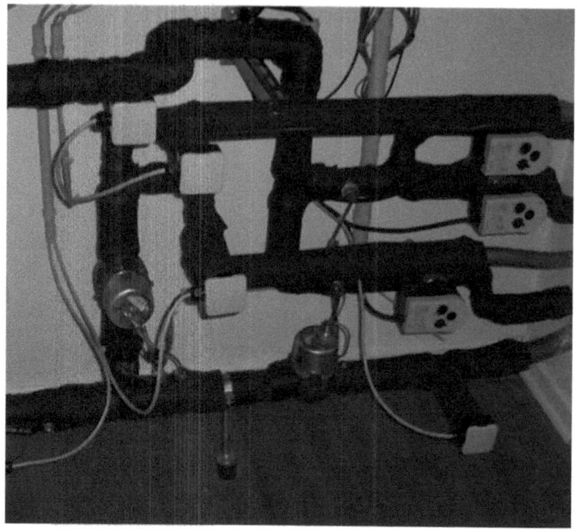

Fig. 10 Measurement of water flow in HVAC subsystems

6.3 Subsystem Consumption Measurement

In order to calibrate enhanced simulation models, the final fuel consumption may not be suitable for the overall HVAC models. Therefore, in these cases, it is also necessary to measure the energy consumption of the HVAC subsystems up to the single components. For this purpose, it can be useful to highlight the mode and the main tools employed for the measurement of thermal and electrical energy flows. These experimental activities are already widely spread in the Anglo-Saxon world as a phase of the commissioning process, during the post-occupancy investigations [22].

In order to quantify the thermal energy flow, the measurement of air/water flow and temperature difference is required (Fig. 10). For example, the heating provided by a radiator is computed starting from the water mass flow rate and the temperature difference between inlet and outlet sections. Therefore, the energy meter already installed in central heating/cooling systems as well as, ad hoc flow meters can be installed for high-rise residential buildings. These electronic energy flow meters offer accuracy up to 1 % as reported in the calibration datasheet.

As regards the measurement of the current in auxiliary systems, the most common instruments are the current transducer (CT). These instruments are placed on wires connected to specific auxiliary system such as motors, pumps or lights and then connected to a digital multi-meter. The CT has typical accuracy up to one per cent. If coupled with the voltage monitoring, by means of a voltmeter, this measure can be used to estimate the appliance power consumptions. However, separate voltage and current measurements should not be used for inductive loads such as motors or magnetic ballasts [27]. In fact, if the signal is distorted owing to various noise sources, a transducer (true RMS) must be adopted. These instruments are capable of accurate measurements of AC voltage even if the waveforms are not purely sinusoidal.

7 Measure of Air and Surface Temperature

The temperature measurement devices commonly used are resistance thermometer (RTD), thermistors and thermocouples. These tools do not directly measure the temperature but rather the variation in a quantity related to the temperature changes. In addition to the different measured variable, these thermometers differ from each other for the cost, the accuracy and the range of measurement. Therefore, the use of a thermometer with respect to another depends on the particular conditions of the temperature to be measured.

The RTD represents one of the most common types of thermometer. In these instruments, the temperature variations are related to the changes in electrical resistance of metal. Platinum RTDs, usually Pt100 or Pt1000, are remarkable instruments: the Pt100 sensor has a resistance of 100 ohm at 0 °C and it is by far the most common type of RTD sensor. These sensors are normally covered by some protective sheath or mounting to form probes that are commonly referred as platinum resistance thermometer (PRT). The tolerances for PRT sensors are specified by the International Standard IEC 751:1983[3]. In this standard, two classes are defined: Class A, with a tolerance of ±0.15 °C at 0 °C; and Class B, with a tolerance of ±0.3 °C at 0 °C. Sometimes the accuracy classes provided by the manufactured are defined as 1/10 DIN or 1/3 DIN. This means, respectively, a certified tolerance of 1/10 or 1/3 of the Class B specification. The linear relation between platinum resistance and temperature, in the range of environmental temperatures, makes them the thermometer of choice for many applications. The main limitation of this type of sensors is connected to the relative higher cost, from 20 to 300 €, linked to the metal cost.

For this reason, RTDs using semiconductors in lieu of metals (i.e., bulk semiconductor sensor or thermistors) are also very popular. Additionally, the semiconductor material exhibits a large change in resistance proportional to a small change in temperature. Besides, thermistors are one of the most accurate types of temperature sensors with a typical accuracy of ±0.1–0.3 °C depending on the particular thermistor model. However, thermistors are fairly limited in their temperature range (typical 0–100 °C) and in the non-linear relation temperature resistance against the lower cost from 0.20 to 20 €.

However, to a considerable extent, the cheaper and widespread thermometer in temperature measures is the thermocouple. This instrument is based on the Seebeck's effect, whereby if two conductors of different material are jointed, a current

[3] DIN IEC 751:1983, Temperature/Resistance Table for Platinum Sensors

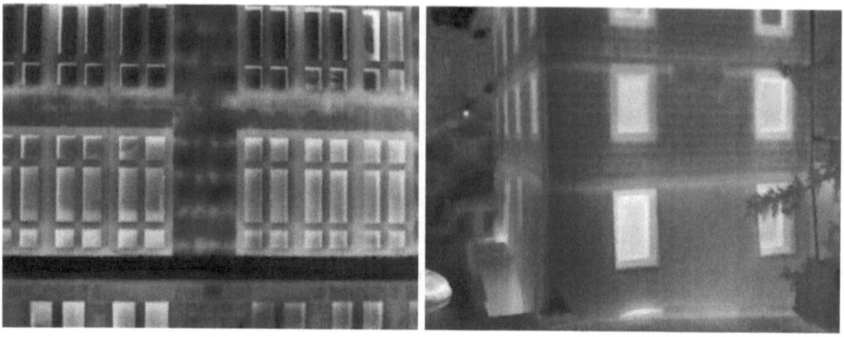

Fig. 11 Thermography analysis of an external wall

grows proportional to the temperature difference of the two joint. A thermocouple is available in different combinations of metals or calibrations. The most common in building analysis are type T for environmental temperature and type *K* for high temperature measurements, such as HVAC systems. Although for the most precise measurements, the reference joint should be kept in a triple point of water, however, such accuracy is rarely needed and a multi-meter reference joint can be adopted.

Nonetheless, the low accuracy frustrates the advantage of the limited cost, from 0.80 to 3 € as a function of the wire insulation and shelter. The EN 60584-2 [25] standard defines the accuracy requirement tolerance that for type *T* thermocouple are within ±0.5 °C for first class and ±1.0 °C for second class.

Starting from the instrument choice, the correct installation position has to be checked, thus ensuring that the boundary conditions do not influence the measurement. For this reason, the instruments must not be placed close the heat sources or direct exposed to sunlight or drafts. Furthermore, for the surface temperature measurements, the position must avoid the edge effects due to the presence of thermal bridges or to non-homogeneous area. For this purpose, the surface should be checked by means of a thermography survey with the purpose of detecting any structural discontinuities or areas with high moisture content (Fig. 11).

8 Building Model Input and Experimental Calibration: Analysis on a Case Study

According to the results of the SA, some parameters were investigated by experimental analysis and more accurate evaluation in order to refine the model.

In particular, the thermophysical properties of the envelope are evaluated both through standard and in situ measurements. The external wall in zone P3_Z1 is 65 cm thick, it has a high thermal mass and it is composed by bricks and sand. Therefore, according to the Italian technical specification UNI TS 11300-1, the

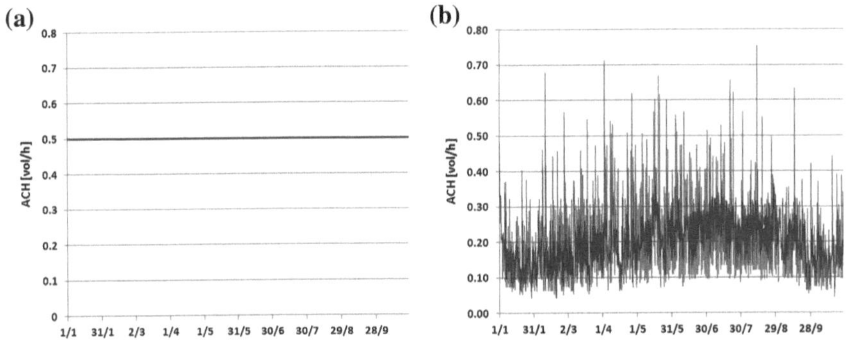

Fig. 12 **a** Standard air-change rate (0.5 vol/h). **b** Calculated air-change rates (EN 15242)

reference structure CO-01 is chosen. Moreover, an experimental analysis is carried out conforming to ISO 9869. Two couples of HFM and thermo-resistance Pt100 are positioned both on internal and external surfaces in order to measure surface temperatures, inward and outward heat fluxes. The measurements are carried out for 70 days (2nd March–10th May) in order ensure the convergence of the solution.

The monitored data are post processed with the average method described in standard ISO 9869. The values of conductance for standard and experimental method are, respectively, equal to 1.372 and 1.552 W/(m^2 K).

Furthermore, considering the absence of HVAC system and the leakages of the envelope, infiltration losses represent a significant contribution. Figure 12a, b show the different air-change rates applied for model definition. Since the envelope presents numerous cracks and leakages, the standard value 0.5 ACH is adopted even if it is used for global natural ventilation. EN 15242 [26] defines a standard method to estimate the infiltration air-change rates, according to envelope features and to local weather data (temperature and wind speed).

In Fig. 13, the instrument position is shown: the heat flux meter (HFM) apparatus (two HFM and two thermoresistance Pt100) is installed in B, while the points from S1 to S5 indicate the thermistors employed for the surface temperature recording. Since the building has no HVAC system, the internal temperatures have been monitored in order to calibrate the simulation model. In particular, both the surface and air temperatures were collected every 10 min in the control thermal zone (i.e., P3_Z1) that is placed on the 4th floor, next to the roof (Fig. 13). The measurement campaign was carried out from March to October 2012.

Starting from the described sources of input data, a series of simulations is carried out with the software TRNSYS (TRNSYS: A Transient Simulation Program, Ver. 16, University of Wisconsin-Madison, 2007). A code identifies each model and it describes which kind of parameter is applied in the analysis. Table 11 reports the set of simulations and it explains which inputs are implemented.

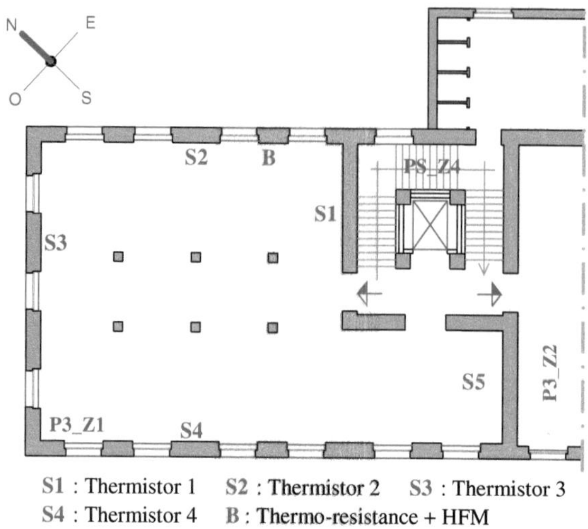

Fig. 13 Control thermal zone: monitoring devices

Table 11 Set of simulations

Simulation	Air-change rates		Thermal conductance	
	0.5	EN 15242	Standard	Measured
tn_05_std	x		x	
tn_05_ms	x			x
tn_en_ms		x		x

9 Validation Indices

9.1 Calibration with Real Consumption

In case of model calibration with real consumption of buildings, for both electrical and fuel energy needs, the reference statistical indices are presented in ASHRAE Guidelines14/2002 [21]. They assess the discrepancies between real and predicted values.

9.1.1 Mean Bias Error $MBE_{\%}$

Mean bias error is evaluated by summing the differences between measured consumptions (*M*) and predicted energy needs (*S*) along the considered time interval and dividing each difference for the corresponding measured value in order to obtain a percentage index.

$$\mathrm{MBE}_{\%} = \left[\frac{\sum_{i=1}^{n}(M_i - S_i)}{\sum_{i=1}^{n}(M_i)}\right] \tag{17}$$

$\mathrm{MBE}_{\%}$ provides for a general assessment of the simulation reliability, in relation to the actual behaviour monitored during the in situ measurement. In addition, it allows to evaluate whether the model overestimates or underestimates the real energy needs. Nevertheless, $\mathrm{MBE}_{\%}$ is not adequate to validate a simulation, because it can give misleading indications due to the sign error compensations. In fact, opposite sign errors tend to neutralize each other and they are not highlighted in the final result. For this reason, additional indices are needed.

9.1.2 Cumulative Variation in Root Mean Square Error

This index is based on the standard deviation between actual (M) and simulated (S) behaviour.

$$\mathrm{RSME}_{\mathrm{period}} = \sqrt{\frac{\sum_{\mathrm{period}}(M - S)^2_{\mathrm{period}}}{N_{\mathrm{period}}}} \tag{18}$$

The Root Mean Square Error determines the absolute value of the discrepancies between two parameters, and it assesses the effectiveness of the simulation in comparison with the real behaviour. The higher the RMSE value, the lower the reliability of the model. However, in case of calibration with real energy consumption, RMSE could be inadequate: in fact the same value of RMSE could be associated both to an accurate model with high consumption and to an inaccurate model with low energy consumption. Therefore, the cumulative variation in RMSE is assumed because it expresses the percentage deviation in relation to a mean value of measured energy needs.

$$\mathrm{CV}\left(\mathrm{RSME}_{\mathrm{period}}\right) = \left[\frac{\mathrm{RMSE}_{\mathrm{period}}}{A_{\mathrm{period}}}\right] \times 100\,\% \tag{19}$$

where A_{period} is an average value based on the number of intervals that characterizes the measurement period:

$$A_{\mathrm{period}} = \left[\frac{\sum\left(M_{\mathrm{period}}\right)}{N_{\mathrm{period}}}\right] \tag{20}$$

9.1.3 Calibration Criteria

The number of measurement intervals depends both on the reference period of the monitoring and on the kind of simulation carried out: in case of steady state

Table 12 Tolerance ranges

Index	IPMVP (%)	M&V (%)	ASHRAE 14 (%)
MBEmonthly	±20	±15	±5
CV (RMSEmonthly)	5	10	15
MBEhourly	–	–	±10
CV(RMSEhourly)	–	–	30

models the calibration is fulfilled through monthly data, while for transient models through hourly or sub-hourly data.

The calibration criteria and the tolerance range have to be established at early stage in relation to the availability of data, the simulation type and the detail level request for the assessment.

The evaluation protocols set different values depending on the model type: for hourly simulations 30 % of discrepancies in terms of CVRMSE and 10 % in terms of MBE are considered acceptable, whereas for the simulations calibrated on monthly data, the tolerance range is halved (Table 12).

9.2 Calibration with Temperature

The model calibration through the temperature monitored in a control thermal zone requires error indices that provide significant indication in terms of temperature discrepancies between predicted and real values.

9.2.1 Mean Bias Error MBE

$$\mathrm{MBE} = \left[\frac{\sum_{i=1}^{n}(M_i - S_i)}{N}\right] \tag{21}$$

The average error MBE is the sum of the differences between measured (*M*) and simulated (*S*) temperatures along the monitoring period divided by the number of records. A positive value of MBE indicates that the model overestimates the temperatures, while a negative value of MBE represents an underestimation of the internal temperature of the control thermal zone. Nevertheless, as highlighted in Sect. 9.1.1, MBE is affected by the sign error compensation, and for this reason, additional indices have to be adopted.

9.2.2 Root Mean Square Error

$$\mathrm{RSME} = \sqrt{\left[\frac{\sum_{i=1}^{n}(M_i - S_i)^2}{N}\right]} \tag{22}$$

Root mean square error RMSE overcomes the misleading of MBE, since it provides for the absolute value of the discrepancies between the temperatures.

9.2.3 Pearson's Index

For model calibration with internal temperatures, the correlation between predicted and real values is a significant indication of the simulation reliability.

The Pearson's index (r) assesses the correlation between two variables, in this case, the temperature trends (predicted and real) and it allows to verify the reliability of simulation:

$$r = \frac{\sum(M \cdot S) - \sum M \cdot \sum^{S}/N}{\sqrt{\left(\sum M^2 - \frac{(\sum M)^2}{N}\right) \cdot \left(\sum S^2 - \frac{(\sum S)^2}{N}\right)}} \tag{23}$$

where:

- M measured temperatures (°C)
- S simulated temperatures (°C)
- N number of intervals

The Pearson's index ranges from −1 to 1: where a negative value means an opposite correlation

- $r < 0$ opposite correlation, if the model temperature increases, the measured values trend to decrease and vice versa; therefore the model is not representative of the real building behaviour;
- $r = 0$ no correlation between variables;
- $r > 0$ direct correlation, if the model and the monitored temperatures have the same trend. $r > 0.5$ represents a significant correlation between temperature variables [3].

The measurement precision can be considered as the reliability limit. This is possible if we consider the correct estimation of the internal temperatures for the control thermal zone aims to validate the simulation energy needs of the analysed building [9].

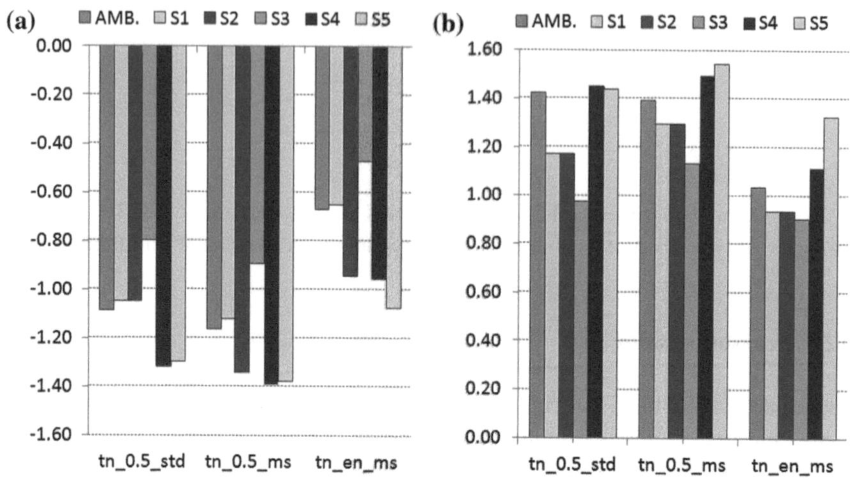

Fig. 14 **a** Mean bias error for hourly temperatures. **b** Root mean square error for hourly temperatures

10 Model Calibration Results

In this section, the calibration process of the dynamic hourly simulation model of the test case (Sect. 3) is presented. The accuracy of the numerical model with respect to the measured data is evaluated both through the statistical indicators (i.e., MBE, RMSE and Pearson's index) and by means of a graphically comparing between the model and experimental trends of temperatures.

The first part of the section will discuss the early stages of calibration. In particular, starting from the initial energy model, firstly the standard values of thermal conductance are replaced with those measured by the in situ tests. Following on from this point, the standard ventilation rate is replaced with the more detailed calculation procedure present in the EN 15242 [26].

In the second part, instead, the final results of the calibration are presented. The calibration is performed by varying the parameters highlighted by the SA according to a uniform distribution, within the plausible range of variability, with the purpose of finding the parameter set that best fits the actual building behaviour. This investigation is carried out on the monitoring period from 2nd March to 22nd October 2012.

Figure 14a and b show the statistical indexes for both air (air) and the envelope surface (S1—S2—S3—S4—S5) temperature in the control thermal zone.

MBE quantifies the long-term performance of a model. A positive value represents the average amount of overestimation in the predicted values and vice versa. MBE in Fig. 14a highlights a general underestimation of the predicted temperature with respect to actual data. Besides, it clearly shows a greater convergence of the refined models with respect to the initial-based model.

Table 13 Pearson's indices—air temperature

tn_0.5_std	tn_0.5_ms	tn_en_ms
0.987	0.991	0.992

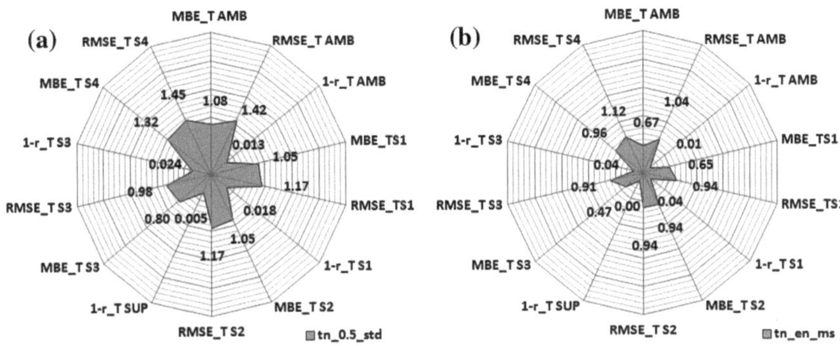

Fig. 15 **a** Error indices for simulation try_05_std. **b** Error indices for simulation tn_en_ms

Nevertheless, MBE drawback arises from the compensation between underestimations and overestimations. Therefore, to assess the reliability of a simulation other indices are required.

In this regard, RMSE overcomes this problem, since it reveals the absolute discrepancies between actual and simulated temperatures. This parameter provides information about the short-term performance of the method by means of a term by term comparison. The smaller the RMSE value, the better are the model provisions. Figure 14b shows a slight convergence of the indexes. In fact, in this test, a few large errors can produce a significant increase in the RMSE index.

These results are also confirmed by the Pearson's index. The values reported in Table 13 highlight as the greatest index increment is obtained using the measured thermal conductance.

Lower increment is achieved by improving the infiltration model. Nevertheless, even Pearson's index shows some weakness. In fact, in case of general uniformity with small deviation between minimum and maximum values, Pearson's index assumes relative high average. Consequently, also this parameter does not permit a univocal assessment of the best model combination. Therefore, a multi-criteria analysis has to be applied by simultaneously plotting the different indexes for all the temperature sensors (Fig. 15). In these pictures, the indexes are plotted for the initial and refined models by means of a radar plots representing the multi-criteria analysis according to the following indices:

- absolute values of mean bias error (|MBE|) for air and surface temperatures (S1, S2, S3, S4, S5)
- values of root mean square error (RMSE)
- complementary values of Pearson's Indices $(1 - r)$.

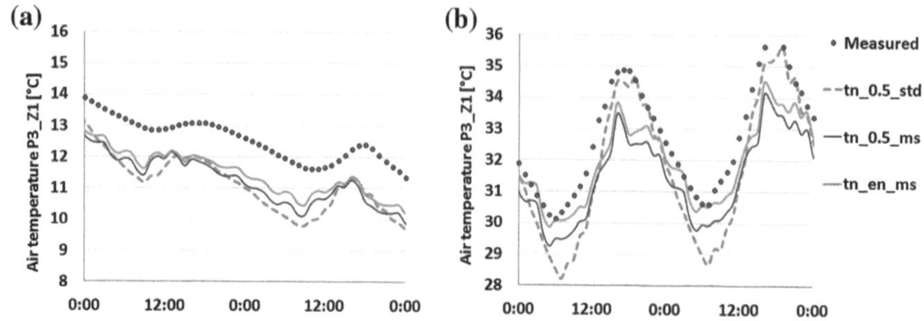

Fig. 16 **a** Air temperature trends (March 4th–5th). **b** Air temperature trends (June 20th–21st)

Table 14 Tolerance ranges

Input	Low limit (%)	High limit (%)
Infiltration rates	−20	+20
U_{roof}	−50	+50
U_{wall}	−20	+20
U_{floor}	−50	+50
Roof capacitance	−50	+50
Wall capacitance	−50	+50
Floor capacitance	−50	+50
g-value for glazing	−30	+30

The higher the blue-painted area, the higher the discrepancies between predicted and actual values of temperatures and, therefore, the lower the reliability of the model. Moreover, this representation highlights the temperature predictions for which the model shows the greatest differences from the measured data.

Nevertheless, error indices give information about the global gap between actual and predicted temperature. A graphical comparison between the hourly trends of temperature is useful in order to have a punctual but qualitative indication of the reliability of building simulations (Fig. 16a, b).

The graphs point out some discrepancies in the evaluation either of positive and negative temperature peaks. Probably, this spread is caused by the incorrect modelling of the thermal capacitance of the walls or to the incorrect estimation of the window solar transmittance.

For this reason, the parameters investigated in the SA are simultaneously varied according to a uniform distribution in order to find the parameter set that best approximates the actual building behaviour. In particular, the range of variation, with respect to the initial value, shown in Table 14 are adopted. These take into account the real knowledge of the parameter and its expected variability.

Several hourly simulations are performed by varying the input data. The evaluation of the model convergence is performed through the multi-criteria analysis previously presented. In Fig. 17a, the internal temperature trend of the control thermal room (measured and simulated values) is reported while in Fig. 17b the radar

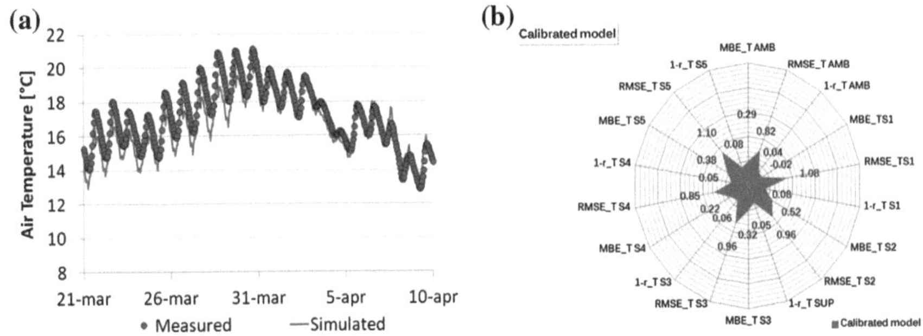

Fig. 17 Calibrated model: **a** Air temperature trends. **b** Multi-criteria radar plot of the calibrated model

graphs obtained for the calibrated model is shown. After the calibration process, the indices MBE and *r* decrease, while RMSE values still remain high. This problem occurs because of the nature of RMSE index, whose amount is affected even by a small number of simulated points that are not consistent with real value. In fact, as it is shown in Fig. 17a, the simulation trend approaches to actual values with lower discrepancies with respect to the previous simulations. Nevertheless, the calibrated model presents some errors in the lower peak values during the winter season.

References

1. Antonacci G, Todeschini I (2013) Derivation of meteorological reference year with hourly interval for Italy. In: Proceedings of 1st IBPSA Italy conference, January 30th–February 1st 2013, University of Bozen, Italy
2. Cappelletti F, Prada A, Romagnoni P, Baggio P (2013) Determination of roof dynamic thermal behaviour by means of in situ measurements: the post-processing analysis. In: Proceedings of CLIMA 2013—11th REHVA world congress and the 8th international conference on IAQVEC June 16–19, 2013, Prague Congress Centre, Czech Republic
3. Giovannini L (2012) Urban scale phenomena and boundary layer processes in mountain valleys. Ph.D. thesis, University of Trento
4. Guan L, Yang J, Bell JM (2007) Cross correlations between weather variables in Australia. Build Environ 42:1054–1070
5. Kioutsioukis I, Tarantola S, Saltelli A, Gatelli D (2004) Uncertainty and global sensitivity analysis of road transport emission estimates. Atmos Environ 38(38):6609–6620
6. Lam J, Hui S (1996) Sensitivity analysis of energy performance of office buildings. Build Environ 31:27–39
7. Lomas KJ, Eppel H (1992) Sensitivity analysis techniques for building thermal simulation programs. Energy Build 19(1):21–44
8. Macdonald IA (2002) Uncertainty in building simulation. Ph.D. dissertation, University of Strathclyde, Glasgow UK
9. Nicholas JV, White DR (2001) Traceable Temperatures. An introduction to temperature measurement and calibration. Wiley&sons, Chicheste
10. Norford LK, Socolow RH, Hsieh ES, Spadaro GV (1994) Two-to-one discrepancy between measured and predicted performance of a 'low-energy' office building: insights from a reconciliation based on the DOE-2 model. Energy Build 21:121–131

11. Ozisik M, Orlande H (2000) Inverse Heat Transfer - Fundamentals and applications. Taylor and Francis, New York
12. Prada A (2012) Energy performance of buildings: modeling of dynamic summer behaviour. Ph.D. thesis, Civil and Environmental Engineering, University of Trento, Trento
13. Raftery P, Keane M, Costa A (2011) Calibrating whole building energy models: detailed case study using hourly measured data. Energy Build 43:3666–3679
14. Raftery P, Keane M, O'Donnell J (2011) Calibrating whole building energy models: an evidence-based methodology. Energy Build 43:2356–2364
15. Scollo S, Tarantola S, Bonadonna C, Coltelli M, Saltelli A (2008) Sensitivity analysis and uncertainty estimation for tephra dispersal models. J Geophys Res B: Solid Earth 113(6)
16. Tian Z, Love JA (2009) Energy performance optimization of radiant slab cooling using building simulation and field measurements. Energy Build 41(3):320–330
17. Tian W, deWilde P (2011) Uncertainty, sensitivity analysis of building performance probabilistic climate projections: a UK case study. Autom Constr 20(8):1096–1109
18. Wilcox S, Marion W (2008) Development of an updated typical meteorological year data set for the United States. In: Proceedings of 37th ASES annual conference, 33rd national passive solar conference, 3rd renewable energy policy and marketing, American Solar Energy Society—SOLAR 2008
19. Pan Y, Huang Z, Wu G, Chen C (2009) The application of building energy simulation and calibration in two high-rise commercial building in Shanghai. In: Proceedings of IBPSA conference, International Building Performance Simulation Association, Glasgow, 2009
20. Yoon JH, Lee EJ (2009) Calibration procedure of energy performance simulation model for a commercial building. In: Proceedings of IBPSA conference, International Building Performance Simulation Association, Glasgow, 2009

Standard References

21. ASHRAE Standard Committee (2002) ASHRAE guideline 14-2002: measurement of energy and demand savings
22. ASHRAE Standard Committee (2005) ASHRAE guideline 0-2005: the commissioning process
23. ASHRAE Standard Committee (2013) ASHRAE handbook fundamentals 2013, ASHRAE
24. CEN Thermal insulation—construction products, building elements and structures. In: Situ measurement of thermal performance—Part 4 testing of structures, (Working group 14)
25. CEN (1993) Thermocouples—Part 2 Tolerances EN 60584-2
26. CEN (2008) Ventilation for buildings—calculation methods for the determination of air flow rates in buildings including infiltration EN 15242:2008
27. "Efficiency Valuation Organisation (2012), International Performance Measurementand Verification Protocol (IPMVP 2012)
28. European Union (2010) Directive 2010/31/EU of the European Parliament and of the council of May 19th, 2010 on the energy performance of buildings (recast). Official Journal of the European Union, 18 June, 2010
29. ISO (1994) Thermal insulation—building elements. In: Situ measurement of thermal resistance and thermal transmittance, ISO 9869:1994
30. UNI (2008) Technical specification, energy performance of buildings. Part 1. Evaluation of energy need for space heating and cooling, UNI/TS 11300-1:2008
31. US Department of Energy (2008) M&V guidelines: measurement and verification for federal energy projects 2008, US Department of Energy. http://mnv.lbl.gov/keyMnVDocs/femp
32. World Meteorological Organization (2008) WMO n. 8 Guide to meteorological instrument and methods of observation

MIX
Papier aus verantwortungsvollen Quellen
Paper from responsible sources
FSC® C105338

FSC
www.fsc.org

If you have any concerns about our products,
you can contact us on
ProductSafety@springernature.com

In case Publisher is established outside the EU,
the EU authorized representative is:
Springer Nature Customer Service Center GmbH
Europaplatz 3, 69115 Heidelberg, Germany

Printed by Libri Plureos GmbH
in Hamburg, Germany